Yamina KORICHI

Pollination of crops by bees

Yamina KORICHI

Pollination of crops by bees

Apoids foraging activity

ScienciaScripts

Imprint

Cover image: www.ingimage.com

This book is a translation from the original published under ISBN 978-620-6-71880-2.

Publisher:
Sciencia Scripts
is a trademark of
Dodo Books Indian Ocean Ltd. and OmniScriptum S.R.L publishing group

120 High Road, East Finchley, London, N2 9ED, United Kingdom
Str. Armeneasca 28/1, office 1, Chisinau MD-2012, Republic of Moldova, Europe
Printed at: see last page
ISBN: 978-620-7-98668-2

THANKS

I thank God for the courage and patience he has shown me throughout my life. route.

*I would like to express my sincere thanks to Ms **AOUAR-SADLI Malika**, Senior Lecturer (A) at Mouloud Mammeri University in Tizi-Ouzou, for agreeing to supervise and direct this work. It is a great honour for me to express my thanks and gratitude to her for her availability, her advice, her guidance, her modesty and for the benevolent interest shown in my work.*

I would also like to thank :

*Mme **MEDJDOUB-BENSAAD F.** professor at the Université Mouloud Mammeri de Tizi-Ouzou for the honour she has done me by agreeing to chair the jury and to give her assessment of this modest work.*

*Mme **SADOUDI ALI- AHMED D.** professeur à l'université Mouloud Mammeri de Tizi-Ouzou et Mme **BENDIFALLAH L.** professeur à l'université M'hamed Bouguerra de Boumerdes pour avoir accepté d'examiner mon travail.*

***Mme MEKIOUS C.** maitres de conférences (A) à l'université Ziane Achour de Djelfa et **Mr ADJLANE N.** maitres de conférences (A) à l'université M'hamed Bouguerra de Boumerdes did me the honour of participating in the examination of this work. I would like to thank them warmly for taking the time to read and examine this thesis.*

*I would also like to thank Ms **GOUCEM-KHELFANE K.** and **Ms CHAOUANE** for their invaluable help.*

*We would also like to thank Professor **Pierre RASMONT** from the University of Mons Hainaut in Belgium for welcoming me to his laboratory on two occasions. We would also like to thank Professor **Denis MICHEZ** from the same laboratory for his help.*

My sincere thanks to the entire Production Laboratory team,

*for their support and encouragement. My warmest thanks to all my friends in the "Plant-animal interaction in natural and cultivated ecosystems" class, in particular **LAMARA MOHAMED R.**, **CHERIFI-HABBI A.** and **IKHLEF H.***

I won't forget all the staff at the Institut Technique Moyen Agricole Spécialisé (I.T.M.A.S) in Boukhalfa for the help and resources provided.

I am particularly pleased to express my deep gratitude and appreciation to my dearest sisters (Nacéra, Zohra and Merzaka) and brothers (Omar, Salah, Mustapha, Mohamed and Aissa), my sisters-in-law, my nieces and nephews for their love, support and encouragement.

Finally, I keep the dearest place in my heart for my mother (God rest her soul) and my father, God bless him, without whom I could never have got this far.

I would like to express my gratitude to all those who have participated in any way in the production of this modest work, not forgetting those who have contributed to our training.

CONTENTS

INTRODUCTION

Pollination is the preliminary act of fertilisation, in which the transfer of pollen between plants enables fertilisation and sexual reproduction. Flowering plants are pollinated mainly by self-pollination, as well as by wind (anemophily) and insects (entomophily). The latter process is the most important type of mutualism between plants and insects. Entomophilic pollination is a production factor in agricultural sectors as varied as fruit growing, seed production and crops under cover.Many flower-feeding insects feed on pollen and/or nectar without being involved in pollination. Pollinating insects include certain Coleoptera, Lepidoptera and Diptera such as Syrphidae, but it is above all the bees (Hymenoptera) that have an inseparable relationship with flowers. In fact, the morphology of bees (the presence of branching hairs on their bodies), their diet (nectar and pollen exclusively) and their foraging behaviour (loyalty to one species of plant during a journey) make them particularly efficient and precise pollen vectors (Michener, 2000). The mutualism that links bees and flowers has led to co-evolution and the diversity of species we know today. More than 20,000 bee species worldwide contribute to the survival and evolution of over 80% of plant species (Vaissière, 2005). The importance of domestic and wild bees in agricultural production is well established, but generally little known. 70% of the world's plant species grown for human consumption depend on entomophilous pollination (Vaissière, 2005). This dependence varies from one crop to another. In fruit crops (almonds, cherries, apples and pears) and some oil crops (rapeseed and sunflower), high yields can only be achieved with the help of insect pollinators. For fodder legumes and many vegetable plants, abundant yields can only be obtained as a result of pollination by bees. This means that if pollinating insects were to disappear, yields would fall significantly. There are also quality aspects to consider: good pollination influences the weight, size and sugar content of the fruit. In fact, bees routinely carry tens of thousands of pollen grains on their bodies and deposit large quantities on the stigmas, resulting in efficient gametic selection of pollen tubes. And in terms of quality, as they move from flower to flower, bees carry pollen from genetically different individuals, and the deposition of allopollen enables cross-fertilisation and reproduction of all self-incompatible species (Vaissière, 2005). The major interest in studying bees is the role they play in natural and agricultural ecosystems. They play a major role in the pollination of a very wide range of wild and cultivated flowering plants (Corbet et al. 1991; Garibaldi et al. 2016). Unfortunately, there is a worldwide decline in many wild bees (Rasmont et al.1993, Rasmont & Iserbyt 2012; Burkle et al.2013; Nieto et al. 2014). Several studies have been carried out on entomophilic pollination to determine the economic importance and impact of insects on crop yields. In Cameroon, Tchuenguem Fohouo et al (2002) showed that bee activity has a positive impact on maize grain yield. This study revealed that the role of Apoids in tropical regions is not limited to entomophilous species but also to anemophilous species (Poaceae, Fagaceae). Borneck and Bricout (1984) and Borneck and Merle (1989), working with the 30 main crops visited by insects in the European Union (EU), determined that entomophilous pollination had an estimated annual value of €5 billion, with €4.3 billion attributable to honeybees. In the United States, bees contribute to the pollination of 130 crops (McGregor, 1976), the annual value of which has been estimated at over 9 billion US dollars (Robinson et al., 1989). Apoids (wild bees and honeybees) have been studied by numerous authors throughout the world in terms of fauna and behaviour. The Maghreb or North Africa has not been the subject of in-depth studies on the apoid fauna. In Algeria, the study of this fauna is very inadequate and fragmentary. Until recently, very little was known about this fauna. The main recent data that have contributed to a better understanding of this group of insects in Algeria are those of Benachour and Louadi (2013), Benarfa et al. (2013), Aguib et al. 2014, Bakiri (2016), Bakiri et al. (2016), Maghni (2017) and Chichoune (2018). These works mainly concern the north-eastern regions of Algeria, namely Constantine. Some other regions of the country have also been the subject of taxonomic studies of bees, such as the work of Bendifallah et al. (2010, 2012, 2013) carried out in the Mitidja region, and that of Bendifallah (2015) in the north-western regions of Algeria. Preliminary inventories have been carried out in a number of Saharan localities by Djouama (2016) and by Cherair (2016) in the Djelfa region. As for the Tizi-Ouzou region, the only studies carried out on wild bees are those by Aouar-Sadli et al (2008, 2012), Aouar-Sadli (2009), Korichi (2015) and Ikhlef (2015). These studies of apoids in Algeria have focused mainly on spontaneous plants and, for the most part, can be summed up as an inventory of the species encountered. The ecology and foraging activity of apoids and the impact of their pollination on crop yields have been given very little consideration. The only work carried out on this aspect is that of Benachour (2008), who studied the pollination activity of bees on 9 market garden plants in Constantine. In 2011, the same author carried out observations on the foraging behaviour of bees on cucumber flowers. Other studies have been carried out on plants such as chickpea (Louadi, 1999), broad bean (Aguib, 2006; Maghni, 2006; Aouar, 2008; Cherair, 2016), potato (Maghni 2006) and tomato (Bendifallah, 2002; Cherair, 2016). Rare observations have been made on fruit trees such as apples (Louadi, 1999), apricots and plums (Cherair, 2016). It is against this backdrop that the present study is intended to help assess the ecological and economic importance of this very useful group of insects.The aim of the present work is to carry out observations on six (06) vegetable plants (broad bean, chickpea, courgette, watermelon, onion and lettuce) and citrus fruits (orange, mandarin and lemon) at Boukhalfa in the Tizi-Ouzou region. During the flowering period of each plant, we will first identify the insect pollinators of the various crops studied by estimating their densities and observing the foraging behaviour and pollination efficiency of the main species. We will also try to understand the foraging behaviour of bees for courgette, broad bean, onion and lettuce. In addition, we will measure the action of entomophilic pollination of broad beans and onions by comparing the yield of plants left free to access these pollinators with caged plants deprived of bees. This thesis is divided into three chapters. The first chapter presents the Apoidea and their mutual relationship with flowering plants. In the same chapter, the various cultivated plants studied are presented. The second chapter describes the study region; in order to facilitate reading of the results discussed below, we also present the equipment and methods used during the experiments carried out at the study station. The third chapter is devoted to the results and discussions of the study of the Apoidea fauna and their relationships with cultivated plants. In this chapter, during the When each plant flowers, we will first identify the pollinating bees and assess the densities and pollination efficiency of the main bee species. We will then measure their action by comparing the yields of plants left free with those of caged plants. The presentation of these results will be accompanied by a discussion that will enable us to highlight the innovations recorded in our study region and a comparison with other work carried out in Algeria and around the world. Finally, we close our report with a general conclusion and a look to the future.

CHAPTER I
BIBLIOGRAPHICAL DATA ON POLLINATION BY APOIDS

1. Overview of Apoids

Insects belonging to the order Hymenoptera encompass more than 100,000 known species worldwide. They are generally characterised by the presence of two pairs of membranous wings, with the front pair being the most developed. These characteristics are described in the root of the word Hymenoptera : Hymenos (membrane), Hymen (god of weddings) and pteron (wing), which relates to the union of the fore and hind wings during flight. The Hymenoptera are divided into two suborders: the Symphyta (which have no wings) and the Hymenoptera (which have no wings).

This group includes the "size group" (the abdomen is largely joined to the thorax, like sawflies) and the Apocrita (characterised by a very narrow "size group", like ichneumons, wasps, ants and bees) (Payette, 1996).

The Apoidea super-family is part of the Hymenoptera, Apocrita and Aculeata (sting-bearing), and includes all bees, whether social, solitary or parasitic. The majority of bees are indigenous, while few species have been introduced or domesticated. Apoids collect nectar and pollen to feed their larvae and are therefore generally referred to as honey bees, thanks to the presence of specific morphological structures that make them exceptionally efficient pollinating insects (Payette, 1996).

There are more than 20,000 bee species worldwide, divided into 9 families (Michener, 2000), 6 of which have been inventoried in Algeria: Colletidae, Andrenidae, Halictidae, Mellitidae, Megachilidae and Apidae (Louadi, 1999; Benarfa, 2005; Maghni, 2006; Bendifallah, 2010; Aouar, 2009; Ikhlef, 2015 and Korichi, 2015). Between each of the six families of apoids, there is a great deal of variation in the length of the mouthparts, particularly the glossae. The first four families (Colletidae, Andrenidae, Halictidae and Melittidae) are often referred to as the "short-tongued" bees, as the majority have a short, broad glosse, whereas bees belonging to the Megachilidae and Apidae families have a well-developed, pointed and often very long glosse (Payette, 1996).

Most pollen-gathering bees have various devices for collecting and transporting pollen. These are mainly harvesting hairs on the hind legs, abdominal brushes or pollen baskets. Kleptoparasitic bee species that do not harvest pollen have no harvesting apparatus. They do not build nests and lay their eggs in the nests of other bees, hence the name "cuckoo bee". The larvae of cleptoparasitic bees do not feed on other larvae. bees but consume the food reserve that is essentially intended for them (Payette, 1996).

1.1. Relationship between plants and bees

Since the Cretaceous period, bees have evolved in parallel with the plants they pollinate. The mouthparts of adults have changed in relation to the shape of the corolla of the flowers they gather. Bees depend exclusively on flowers for their food. They use nectar, which is rich in sugar and water, and pollen, which provides other essential nutrients (Louveaux, 1980). Through their special and important relationship with flora, bees play a vital role in maintaining vegetation in temperate and tropical regions.

Studies by Chansigaud (1975), Batra (1984), Abrol (1988) and Jacob-Remacle (1989 a and b, 1990) show that wild bees play a remarkable role in the pollination of crops and spontaneous plants. Over the last decade, the development of controlled pollination techniques has focused attention on apoids as auxiliaries in agriculture and arboriculture.
Following the domestication of the Apis genus in ancient times, man has now succeeded in domesticating several species of wild bee. The first species was discovered for pollination in 1930 in Japan by Matsuyama. It is a small brown bee, Osmia cornifrons Radoskowski. This bee is 80 times more efficient than an Apis mellifera worker bee. According to Maeta and Kitamura (1981), cited by Pesson and Louveaux (1984), it pollinates a third of Japanese apple trees. The Belgian company Biobest has bred the Bombus terrestris bumblebee for use in pollinating greenhouse and field crops. In Canada and the United States, farmers have tamed Megachile rotundata Fabricius, 1787 (Megachilidae) and Nomia melenderi Cockerell (Halictidae) by creating nest boxes for them around farmland (Batra, 1984).

2. Pollination

Pollination is a key stage in the reproduction of angiosperm plants (70 Between % and 90% of angiosperms are pollinated by an animal species). This is the process of transferring the pollen produced by the anthers of the stamens to the stigma of the flower, followed in principle by fertilisation of the ovule by the pollen grain germinating on the stigma. Pollination of a plant by insects depends on floral morphology, the degree of self-fertility and the arrangement of flowers on the plant or on neighbouring plants. In this sense, plants can be divided into two categories:

- Self-pollinating plants, whose flowers are pollinated by their own pollen or, by extension, by pollen from another flower on the same plant. However, bees often optimise pollination even in self-pollinating flowers. Pollination in other plants, particularly cereals, is carried out by wind and gravity, with bees playing only a minor role.
- Allogamous plants, whose flowers cannot be fertilised by pollen produced by the plant's various flowers. They require cross-pollination, which ensures better genetic mixing than self-pollination.

Cross-pollination is mandatory in :

- unisexual plants ;
- bisexual plants, which are self-sterile (the pollen from one plant cannot be used for other plants). germinate on the stigmas of the same plant).
- Heterostylous plants, in the case of the primrose, have two types of flowers in roughly equal proportions: longistyles (long style) and brevistyles (short style). Pollen from longstyles can only germinate on the stigma of brevistyles and vice versa.
- hermaphroditic plants, known as dichogamous plants, which show a difference in the maturity of their sexual organs.

The presence of a pollen vector is therefore essential for cross-pollinating plants, but it can also be very beneficial in certain self-pollinating plants, by allowing the most complete fertilisation possible (Jacobe-remale, 1990).

2.1. Impact of cross-pollination on fruit and seeds

Cross-pollination and self-pollination are two methods of reproduction which do not have the same genetic value, and the superiority of the heterozygous state over the homozygous state has been demonstrated by several studies. Trials carried out in cages with and without bees by Barbier (1986) on oilseed rape and apple have shown that the fall in young fruit is greater with self-pollination, whereas with cross-pollination there is no fall at all. As for fruit development, the same author reported that in apple, the average weight of ovaries, which was 40 mg on the day of pollination, reached 292.4 mg 14 days after cross-pollination compared with 212.8 mg the same number of days after self-pollination.

In the case of self-pollination, the small number of seeds can result in deformed fruit of lesser commercial value. This is the case with certain pear varieties which, due to a lack of seeds, produce smaller, asymmetrical fruit. As for With regard to the germinative quality of the seeds, several authors have shown that the energy and germinative power of seeds obtained by cross-pollination are clearly superior to those of seeds formed by self-pollination, which gives them a high commercial and agricultural value (Barbier, 1986). Come (1968), quoted by Barbier (1986), was able to demonstrate the direct relationship between the number of seeds per apple and shelf life: the higher the number of seeds, the less quickly the apple wilts and the longer it can be kept, and he found that it was the seeds with high germinative energy that enhanced the keeping qualities of apples.

2.2. Factors influencing pollination entomophilic

Certain physical environmental factors influence pollinator activity and therefore pollination, such as light, temperature, humidity and wind. Light plays a part in the perception of the difference between day and night. Some insects are diurnal, while others are nocturnal or crepuscular. The activity of certain pollinating insects depends on the perceived light. Light can be modified by the presence of clouds or thunderstorms, which will favour the activity of crepuscular insects. The effect of temperature is fairly closely linked to that of light, and is more of a seasonal effect. The lowest temperature at which honeybees can actively fly is 10°C (Kevan and Baker, 1983 cited by Dibos, 2011). Humidity has an indirect impact on pollination, since this factor does not affect pollinator activity, but rather the sugar concentration of the nectar. Lastly, wind disrupts flight, while rain prevents any bee activity (Kevan and Baker, 1983 cited by Dibos, 2011).

2.3. Bees and pollination

Bees and flowers are inseparable, and the mutualism that links them has led to co-evolution and the diversity of species we know today (Vaissière, 2002). Bees and flowers both have a vital need for each other in order to live. Indeed, the pollination of flowering plants is largely carried out by bees, which depend entirely on flowers for their food: adults consume nectar, while larvae feed on pollen with added nectar (Jacob-Remacle, 1990). When visiting flowers, these insects routinely carry tens of thousands of pollen grains on their bodies and deposit large quantities on the stigmas, resulting in efficient gametic selection of pollen tubes (Fig. 1). In terms of quality, as they move from flower to flower, bees carry pollen from genetically different individuals, and the deposition of allopollen enables cross-fertilisation and reproduction of all self-incompatible species (Vaissière, 2005).

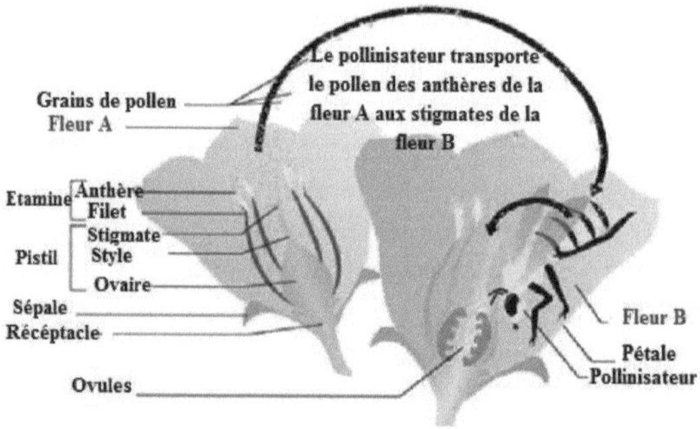

Figure 1: Bee pollination process (Anonymous 1, 2013)

To attract bees, plants have developed floral organs, some of which are highly complex. As well as offering nectar, they are often brightly coloured to attract insects, and have fragrances whose molecules can be perceived by bees up to 1,200 metres away. The most remarkable initiatives undoubtedly come from certain orchids whose flowers imitate the morphology of the female of a particular species of bee and produce odorous substances that counterfeit the pheromones of those of the female of this insect. The lured male will mate with the flower (pseudo-copulation) and transport its pollen, once again fooled, to the next orchid of the same species (Lachaud and Mahe, 2008).

Apoids have also developed morphological and ethological adaptations for harvesting nectar or pollen (Tasei, 1984). These insects have modified mouthparts and, in particular, an elongated tongue with a gutter-like tip for licking and sucking nectar. Likewise, they have devices (tegumentary pilosity or appendages) and a particular morphology of the legs, which facilitate the collection and transport of pollen (Jean-Marie, 1991). The activity of native Apoids is characterised by short foraging periods synchronised with the flowering period of their host plants, and the majority are spring species (Payette, 2003). According to Jacob-Remacle (1990), bees can be classified into three categories according to their specific dietary requirements for pollen:

- **polylectic** species, which are the most numerous, obtain their pollen from an a large number of plants belonging to various families;

- **oligolectic** species collect pollen from a group of plants belonging to a family;

- monolectic species exploit a single genus or even a single species floral.

Overall, however, both male and female bees were more eclectic in their choice of flowers for nectar than in their choice of flowers for pollen.

2.4. Use of wild bees in the pollination of cultivated plants

Studies show that wild bees and other wild pollinators play a key role in the development of our environment. They play a key role in the pollination of both wild and cultivated plants.

• Some wild bees have a lower thermal activity threshold than domestic bees. Bumblebees are remarkable in this respect: they can be observed in plantations in very cool weather, even in the rain. Their activity and foraging behaviour are constant from one year to the next; during flower visits, the frequency of contact with stigmas is very high.

• The foraging speed of certain genera (Osmies, Anthophores, Bumblebees) is

than that of the honey bee.

• Solitary Apoidea females carry a larger quantity of pollen grains on their bodies than do honey bees; moreover, unlike social Apoides, they do not moisten the pollen they accumulate in their collecting apparatus; this pollen can therefore be detached more easily when the female is placed on the flower's reproductive organs (Jacobe-Remacle, 1990). These wild bees are currently being bred in some countries to pollinate crops that are poorly pollinated by honey bees. Insufficient or inadequate pollination by the domestic bee may be due to poor adaptation to the morphology of the flower (e.g. alfalfa) or to reduced activity during the summer. very early flowering at low temperatures (e.g. almond tree) (Benachour, 2008). Farmed wild bees include :

• The leafcutter bee Megachile rotundata is the subject of a veritable industry and intense trade in tens of millions of cocoons in North America. Canada is the biggest producer and exporter of cocoons. In Europe, a number of countries have also embarked on cocoon farming, mainly France, Spain and certain Eastern European countries. This bee is mainly used for pollinating alfalfa.

• Bumblebees are reared for greenhouse pollination of crops such as tomatoes, aubergines and peppers. The morphological and behavioural characteristics of bumblebees mean that they are superior to honeybees in pollinating these crops (Pouvreau, 2004).

• Osmies (Megachilidae) bred mainly for the pollination of fruit crops in the USA, Japan and Spain. In Canada and the United States, we find Osmia lignaria, a bee native to this region, which is of great interest in the pollination of fruit trees and is easy to manage due to its favourable biological characteristics. The mason bee Osmia cornuta, with a few hundred individuals sufficient to pollinate a hectare of apple or almond trees, where several thousand domestic bees would be needed. In Japan, Osmia cornifrons i s bred and marketed for pollinating apple orchards (Bosch & Kemp, 2002 cited by Benachour, 2008). This bee is able to operate in a wide range of ambient temperatures and has an individual pollination efficiency 80 times higher than that of the honey bee (Jacob-Remacle 1990).

2.5. Impact and economic impact of Apoidea pollination on agricultural production

Bees help to pollinate plants, which has a positive impact on the environment. They are involved in almost 80% of the pollination of plant species and are therefore an essential link in the survival, evolution and reproduction of plants. Bees play a crucial role in food security and nutrition, with a third of our diet directly or indirectly dependent on the activity of these insects. According to the Intergovernmental Science-Policy Platform on Biodiversity and Ecosystem Services, the value of the ecological and economic services provided by bees corresponds to 577 billion dollars. Through the pollination of plants, bees promote agricultural production, which ensures food security, and through their high nutritional value products (honey, royal jelly, pollen, etc.), ensures the nutritional security of the population

(Fao, 2019). These insects are involved in the pollination of ¾ of cultivated plant species in four main sectors (Vaissière, 2005):

- Fruit growing, particularly rosaceous fruit trees (apricot, almond...), cherry, peach, pear, apple and plum) and kiwi;

- Field crops: cereals (buckwheat), oilseeds (rapeseed and sunflower) and protein crops (faba beans);
- market garden crops: cucurbits (courgettes, melons and watermelons), solanaceous plants (tomatoes, peppers and aubergines), rosaceous plants (strawberries and red berries);
- Forage species such as alfalfa and clover, and other species such as nuts and stimulating spices (cocoa, coffee, etc.) are also dependent on pollination.

3. Agriculture in Algeria

Algeria has a highly diversified natural environment, but in terms of agriculture, it is not self-sufficient and has a significant deficit every year, with almost 75% of its needs being imported. Since the 2000s, the country has implemented a policy aimed at improving national food security, developing certain priority agricultural sectors and developing land. This policy has been implemented through various national plans shaped by fluctuations in oil revenues, including the National Agricultural Development Programme (PNDA 2000-2010), the Agricultural and Rural Renewal Policy (PRAR 2010-2014) and now the FELAHA Plan 2014-2020 (Bessaoud, 2019). Algeria is the largest country in Africa, with a surface area of 2,381,741 km^2. Arable land covers 8.5 million ha, or 3.5% of the total surface area. In 2017, more than half of the UAA (utilised agricultural area) was devoted to arable crops, particularly cereals. These crops dominate in the semi-arid zones of the high plains of the Tellier region, the sub-arid zones of the high plateaux, and the humid and sub-humid zones of the coastal and sub-coastal regions.
Fruit growing (just over 10% of the UAA) is organised around olive, date palm and fruit tree plantations. Market gardening, which covers approximately 5% of the UAA, is practised in coastal and sublittoral areas. and inland plains. The main market-gardening wilayas are Boumerdes, Ain Defla, Biskra, Algiers, Mascara, Mostaganem, Skikda, Tipaza, El Taref, Tlemcen, M'sila and Chlef. The staple vegetables are potatoes, tomatoes, garlic, onions, green beans, carrots, turnips and peppers. Farmers in the oases of the south maintain gardens co-planted with date palms, market garden produce and fruit trees, irrigated in some regions using techniques developed as early as the 11th century (foggaras) (Bessaoud, 2019).

3.1. Market gardening in the Tizi-Ouzou region

Vegetable production in the wilaya of Tizi-Ouzou includes cereals, fodder crops, pulses, market gardening, fruit growing and viticulture. Useful agricultural area (UAA) increased from 94,152 ha in 1996/1999 to 96,414 ha in 2000/2006. Irrigated agricultural land also increased from 4,589 ha in 1996/1999 to 5,128 ha in 2000/2006. Cereal production includes durum wheat, common wheat, barley and oats, but the dominant cereal crop in the wilaya of Tizi-Ouzou is durum wheat. The average area under cereals fell from 12,838 ha in 1996/1999 to 9,714 ha in 2000/2006. Market garden crops include potatoes, watermelons, melons, onions, tomatoes, carrots, broad beans and others, with an average annual area of

8,089 ha during the period 1996/1999 and 7,200 ha during the period 2000/2006, but the dominant market garden crop in the wilaya is potatoes (Chedded, 2015).

4. General information on the plants studied

This section gives some general information on the plants studied, their economic importance in Algeria and a description of their inflorescence.

4.1. Chickpea (Cicer arietinum)

Chickpea (Cicer arietinum L.) is a plant in the Fabaceae family, probably originating in the Near East, where three wild annual species still exist. Chickpeas are mainly grown in Mediterranean regions, southern Asia and South America and play an important role in the human diet. This legume is ranked in 14ième order in terms of the surface area it occupies and comes in 16ième order in world production estimated at around 7 million tonnes for a surface area of 1 million hectares (Knights et al., 2007 cited by Bouri and Gaouar, 2018). In Algeria, chickpeas play an important role in our diet. As a result, current production is very low and the surface area occupied by this species is gradually decreasing (Abdelguerfi et al., 2001a cited by Bouri and Gaouar, 2018), from 19290 in 1994 to 4600 in 2008 (Anonymous 2, 2009).

4.1.1. Position systematically

According to Crété (1965) the systematic position of the chickpea is as follows: Kingdom: Plant
Phylum: Phanerogams (Spermaphytes)
Subphylum : Angiosperms
Class : Broadleaf plants

Subclass : Dialypetal

Order : Leguminous plants (Fabales)

Family:Legumes (Fabaceae)

Subfamily : Papilionaceae

Tribe :Victims

Genre :Cicer

Species :Cicer arietinum

4.1.2. The flower

The chickpea flower has a tube-shaped calyx surrounding the bud with five pointed lobes (sepals) open at the tip. The corolla has five petals (the standard, wings and carina). The chickpea has ten stamens, nine of which form a tube around the ovary. The upper part of the threads is free and turned upwards, and each supports an anther. The uppermost stamen is free and separate from the others. The stigma is a swelling at the end of the style, facing upwards like the filaments. The ovary contains ovules that develop into seeds after fertilisation (Fig. 2).

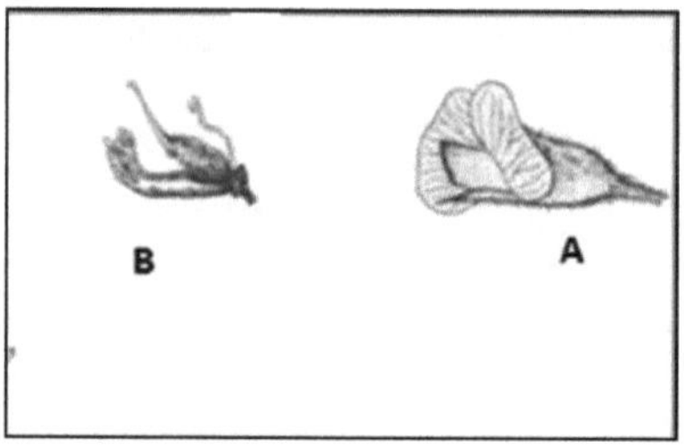

Figure 2: Morphology of the chickpea flower
(A) Zygomorphic flower, (B) stamens, pistil and ovary (Blumber., 1991).

4.2. Fava beans (Vicia faba L.)

Broad bean (V. faba L.) is an important legume grown on around three million hectares worldwide (Link et al., 1999), with more than 50% of the cultivated area in China, 20% in North Africa and 10% in Europe (Abu-Amer et al., 2011). Its cultivation in the countries of the Mediterranean Basin accounts for almost 25% of the total cultivated area and world bean production, with a yield very close to the world average (Saxena, 1991). In Algeria, bean cultivation is mainly concentrated on coastal plains and sublittoral areas, with a cultivated area of around 65,000 ha and production of between 20,000 and 38,000 tonnes per year (Zaghouane, 1991cité par Benachour, 2008). In the wilaya of Tizi-Ouzou, the average area set aside for bean cultivation is 821.1 ha, which varies from year to year, influencing production, which also varies, with a ten-year average (2001-2012) of 9040.6 qx (Khelloul, 2014).

4.2.1. Position systematically

According to Reta Sanchez et al (2008), the broad bean is botanically classified as follows:
Kingdom:Plant
Division :Magnoliophyta

Class :Magnoliopsida

Order :Fabales

Family :Fabaceae

Subfamily :Faboideae

Tribe :Vicieae

Genre :Vicia

Species :Vicia faba L.

4.2.2. Inflorescence

A bean plant bears 50 to 80 flowers grouped in clusters of 2 to 9 flowers. Each of these inflorescences starts in the axil of a leaf. Flowering starts at the bottom of the stem and progresses upwards, tier by tier. The flower is typical of the Papillionaceae, with a large corolla measuring 25 mm in length, coloured pinkish-white with black spots. It consists of a large petal (the standard), two lateral petals or wings covering two other lower petals joined along their outer edges to form the carina. There are 10 stamens, 9 of which are fused together by their threads in a sheath (Fig.3).

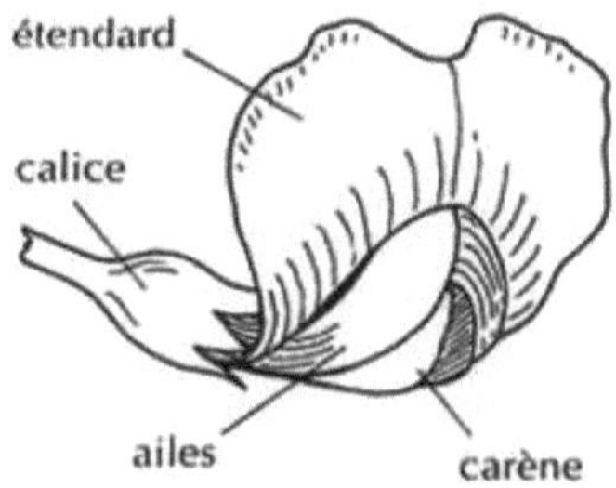

Figure 3: Diagram of the V. faba flower (Jaccon, 2015)

4.3. Courgette (Cucurbita pepo L.)

Courgette (Cucurbita pepo L.) is an annual plant belonging to the Cucurbitaceae family. It is native to America and was widely cultivated in Central and South America. This species may have been domesticated at least twice and in two different regions: in Mexico and the eastern United States, C. fraterna and C. texana are possible progenitors (Dhaliwal, 2017). At present, they are widely cultivated in every country in the world where they find favourable conditions for their development. This vast spread of squash is due to the nutritional and taste qualities of the fruit. They are most widely grown in the Mediterranean region, Asia and the United States. China is the biggest producer of summer squash, followed by Ukraine, Afghanistan and Turkey. Together, they produce 45% of total world production (Dhaliwal, 2017). In Algeria, it is one of the most widely grown crops, thanks to the favourable climatic conditions and soil type. Production exceeds 227,000 tonnes and has doubled since 2000. Production areas include Mostaganem, Algiers, Boumerdes, M'Sila and Tipaza. Tunnel crops in Tipaza, Biskra, Algiers and Mostaganem account for production of 33,000 tonnes (Anonymous 3, 2014).

4.3.1. Position systematically

According to Feller et al (1995) the courgette is scientifically classified as follows: Kingdom: Plantae
Division :Magnoliophyta
Class :Magnoliopsida

Order :Violales
Family :Cucurbitaceae
Genus :Cucurbita
Species :Cucurbita pepo L.

4.3.2. Flower

Courgette is a monoecious plant, with each plant bearing two types of flower: male and female. The male flowers are always more numerous than the female flowers, which is necessary to ensure that the latter are fertilised.
Squash flowers are very large, orange-yellow in colour and unisexual, male or female. The male flowers have 5 stamens with sinuous anthers. The female flowers can be identified by the small crown on the underside of the flower, which corresponds to the pistil. Squash flowers have a short lifespan. They open in the morning and close around midday, never to open again. Each female flower therefore has a few hours in which to be fertilised. Male flowers can be recognised by their long, slender stalk and the fact that they have no inferior ovary beneath the flower and an androecium inside, whereas females have an inferior ovary appearing before the flower and a pistil inside (Fig. 4).
Male flowers produce nectar and pollen, the grains of which are large and well-suited to being carried by insects, while female flowers produce only nectar and in larger quantities, which means they attract a greater number of bees (Nepi and Pacini, 1993).

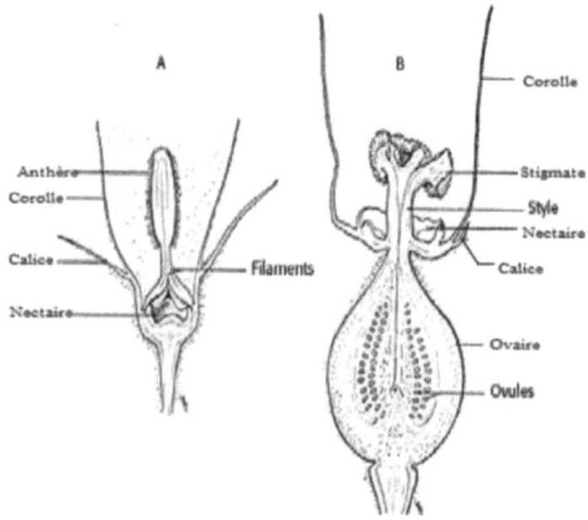

Figure 4: Longitudinal section of male (A) and female (B) squash flowers (Mc Gregor, 1976).

4.5. Watermelon (Citrullus lanatus)

Watermelon (Citrullus lanatus) is an annual herbaceous plant in the Cucurbitaceae family, and is monoecious, bearing both male and female flowers. It is one of the most important vegetable crops in the Cucurbitaceae family, and is widely grown throughout the world. It originated in the dry regions of southern Africa (Wein 1997 cited by Mc Gregor, 1976). It was first cultivated in hot, dry, tropical and Mediterranean countries, before being introduced into hot, humid regions (Pitrat et al, 1999 cited by Hamza, 2014). World watermelon production for 2011 was 104,472,354 tonnes, in Algeria it was 1,285,134 tonnes. According to the FAO (2011), the five largest watermelon producers in the world are China, Turkey, Iran, Brazil and the United States (Hamza, 2014).

4.5.1. Position systematically

According to Wehner (2008), the botanical classification is as follows:

Kingdom :Plantae
Under reign :Tracheobionta

Division :Magnoliophyta

Class :Magnoliopsida

Subclass :Dileniidae

Order :Violales

Family :Cucurbitaceae

Genus :Citrillus

Species :Citrillus lanatus Thunb.

4.5.2. The flower

Watermelon flowers are small with pale yellow corollas and appear in the axils of the leaves. leaves, they are, as in most Cucurbitaceae, either male or female, but in this case theAll are present on the same plant (monoecious plant) (Fig.5). The flowers open 2 cm from the base.hours after sunrise. Pollen is generally released before the flower opens, but remains on the anthers in sticky masses unless removed by insects. The stigmas are receptive as long as the flower is open, but maximum receptivity occurs between 9 and 10 o'clock (Adlerz, 1966 cited by Delaplane and Mayer, 2000). The flowers close in the afternoon and never reopen, even if they are not pollinated. Watermelon flowers attract bees for nectar and pollen, but their density is low and the bees are easily distracted in favour of richer resources (Delaplane and Mayer, 2000).

Figure 5. Photograph of male (A) and female (B) Citrullus lanatus flowers. (Original photograph, 2016).

4.5. Onion (Allium cepa L.)

Originating in Central Asia, the onion has been cultivated since ancient times. It is one of the most widely cultivated crops (Schweisguth, 1976), ranking fourth in the world in terms of vegetable production. In Algeria, onion production has more than tripled in 10 years, reaching 1,183,268 tonnes in 2012. Onion production is particularly high in Mascara (200,000 tonnes), Tiaret (174,000 tonnes), Mascara (200,000 tonnes) and Tiaret (174,000 tonnes). Boumerdes (140,000 tonnes) and Skikda (109,000 tonnes) (Anonymous 3, 2014). Allium cepa L. is a biennial Liliaceae, producing an edible bulb in the first year and a spherical inflorescence at the end of the second year. The small white flowers are followed by capsules containing several seeds.

4.5.1. Systematic position

According to Nyakabwa (2009), the systematic position of the onion is as follows:

Kingdom :Plantae
Under reign :Tracheobionta

Division :Magnoliophyta

Class :Magnoliopsida

Subclass: Liliadae

Order :Liliale

Family :Liliaceae

Type :Allium

Species :Allium cepa L.

4.5.3. Inflorescence

Each inflorescence (umbel) consists of 50 to 2,000 greenish-white flowers. Each flower, 3 to 4 mm long, has 6 stamens arranged in 2 whorls, and a style with a trilocular ovary, with 2 ovules per cell. The anthers of the inner stamens open first, releasing their pollen one after the other; then the anthers of the outer stamens open in turn, at equally irregular intervals. The onion is an allogamous species, so the male reproductive organs release pollen before the female reproductive organs are receptive (Fig.6). In fact, the anthers of a flower open first and release pollen for 3 days before the stigma of the same flower is receptive for 6 days. The flowers bloom for 4 weeks, peaking in the second week. Most of the pollen is released between 9 and 5pm on the first day of flower opening, before the stigma is receptive. The nectaries are located at the base of the stamens and the nectar accumulates between the ovary and the inner stamens (Pesson and Louveaux, 1984).

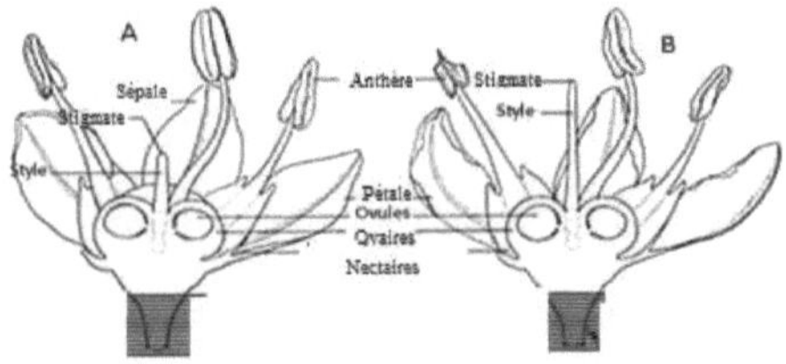

Figure 6: Onion (Allium cepa) flower (A) sagittal section in pollen production phase, (B) flower in stigmatic receptivity phase (Mc Gregoor, 1976).

4.6. Lettuce (Lactuca sativa)

Lactuca sativa is an annual herbaceous plant in the Asteraceae family. Lettuce has two very distinct phases: the vegetative phase, forming a more or less closed head, corresponds to the stage of development used for marketing, and the reproductive phase during which the main stem elongates (bolting) leads to flowering and seed production. According to Boissier, the lettuce species may have originated from wild forms in the mountains of Kurdistan. World production of lettuce is around 22 million tonnes. The main producing countries are China (10 million tonnes), the United States (5 million tonnes), Spain (1 million tonnes) and Italy (0.83 million tonnes), with France producing 0.46 million tonnes (Anonymous 3, 2017). In Algeria, it is grown in all areas, with the most widely cultivated varieties falling into the following groups:

- Cutting lettuce: blond lettuce, American curly lettuce
- Head lettuce: Queen of May, golden yellow gotte, batavia, 4-season wonder,

stubborn from Nîmes, divina.

- Romaine lettuce: Balen, blond market garden (Anonymous 4, 2010).

4.6.1. Position systematically

According to Chase and Revel (2009), lettuce is classified as follows:

Kingdom :Plantae
Under reign :Tracheobionta

Division :Magnoliophyta

Class :Equisetopsida

Subclass :Magnoliidae

Order :Asterales
Family :Asteraceae
Subfamily :Cichorioideae
Tribe :Lactuceae
Sub-tribe :Lactucinae

Genre :Lactuca
Species :Lactuca sativa L.

4.6.2. Inflorescence

The inflorescence is a panicle made up of several clusters. Each cluster bears 15 to 25 or more flowers. The pollination mechanism is aided by self-pollination, with the bracts beginning to open as the flowers develop. The flowers elongate rapidly 24 hours before anthesis. The style and stigma are covered in white bristles. Anther dehiscence occurs before pistil elongation. As the pistil elongates, the bristles next to it push the pollen out of the pollen sac. The lobes of the stigma separate, allowing the pollen to fall onto the stigmatic papilla, located on the inner surface. This is how fertilisation takes place (Salunkhe et al 1987 cited by Hebie, 2001).

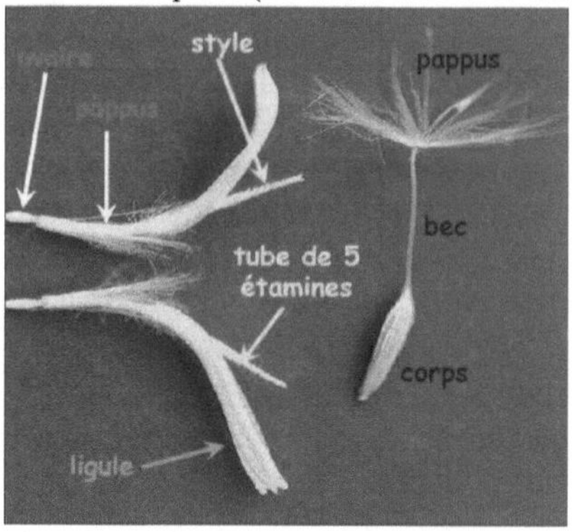

Figure 7: Photograph of the morphological appearance of a floret and an achene containing the seed of lettuce (Zorrig, 2011)

4.7. Orange

The orange tree (Citrus sinensis L.) is a species of fruit bush in the Rutaceae family, grown in warm regions such as the Mediterranean. This ancient hybrid is probably a cross between the grapefruit and the mandarin. The cultivation of the orange tree goes back a very long way, and is intertwined with the history of China, where it originated. During the first millennium BC, the orange tree spread rapidly throughout South-East Asia, arriving in the Mediterranean in the 7th century. World citrus fruit production reached 60 million tonnes in 2000, of which Brazil is the largest producer with 18 million tonnes, followed by Florida with 11 million tonnes and the Mediterranean basin with 10 million tonnes. (Anonymous 5, 2012). Algeria's citrus orchards (mainly oranges, with a few clementines and lemons and very few mandarins and pomelos) are estimated at 65,000 ha, over 50% of which are located in the Mitidja region. The wilayas of Annaba, Skikda, Oran, Mascara, Mostaganem, Chlef, Blida, Algiers and Tipasa are the main producing areas (in total, 43% of citrus fruits are grown in the Mitidja plain, 27% in the Chélif region and 7% in Mascara). The orchards are located on rich, irrigated land (Anonymous 3, 2014).

4.7.1. Position systematically

According to Praloran (1971) the taxonomic position of citrus according to Swingle is as follows:

4.7.2. The flower

Kingdom :Plant
Phylum :Angiosperms
Class :Eudicotes
Subclass :Archichlomydeae
Order :Germinal(Rutales)
Family :Rutaceae

Subfamily :Aurantioideae

Tribus :Citreae
Sub-tribe :Citrinae

Genre :Citrus

Species :Citrus sinensis L.

The flowers are between 2 and 4 cm in diameter, axillary, fragrant, simple, often perfect (with functional stamens and pistils) and sometimes staminate (with stamens but no pistils). The calyx is lobed, generally made up of five petals with glands containing aromatic essences. There are between 20 and 40 stamens and the petals are pinkish-white, pinkish-purple towards the outside in lemons and reddish in other citrus varieties (Fig.8). The ovary is surmounted by a massed stigma, composed of 8 to 18 locules (cavities), with 4 to 8 ovules per locule, in two rows (Selka, 2007).

Figure 8: The orange blossom (Original photograph, 2016)

CHAPTER II
MATERIALS AND METHODS

1. Introduction to the region

1.1. Geographical data for the wilaya of Tizi- Ouzou

The Tizi-Ouzou region (Greater Kabylie) is located on Algeria's central-eastern coastline and covers an area of 2,958 km^2 . It lies between 3°35'and 5°05'E longitude, and between 36°22'and 36°55'N latitude. It is bordered to the north by the Mediterranean Sea, to the south by the Wilaya of Bouira, to the east by the Wilaya of Bejaia and to the west by the Wilaya of Boumerdès. Its geographical relief is characterised by the vast mountainous regions of the Djurdjura, which peaks at an altitude of 2,308 m, a coastal range of high hills at an altitude of 500 to 1,000 m, and the Sébaou valley, whose altitude barely exceeds 500 m. This valley is meandered by the Oued Sébaou, which is the main source of water (Belhadi, 2010).

1.2. Climate data

The Tizi-Ouzou region, in northern Algeria, is influenced by the Mediterranean climate. The climatic data used for the analysis are from the "Office National Météorologique" weather station in Tizi-Ouzou.

1.2.1. Temperature

Temperature is a very important element of climate. It is linked to solar radiation and altitude and local environmental conditions. The temperatures recorded in 2016, 2017 and 2018 (Tab. 1) show that the coldest month during this period was March, January and February respectively, with average monthly temperatures of 12.5°C, 8.9°C and 9.9°C. The warmest months were July in 2016 and 2018, and August in 2017, with mean monthly temperatures of 27.9°C and 29.7°C respectively.

Table 01: Average monthly temperatures (°C) at the Tizi-Ouzoude weather station 2016 to 2018.

	Jan	Feb	March	Apr	May	June	July	August	Seven	Oct	Nov	Dec
2016	12.8	12.6	12.5	16.1	19	24.3	27.9	27.1	24.3	22.5	15.7	12.6
2017	8.9	12.7	14.5	16.4	21.8	26.4	29.4	29.7	23.9	19.5	13.7	10.4
2018	11.2	9.9	13.6	16.5	17.8	22.8	27.8	27.4	25.3	19.1	14.9	11.9

1.2.2. Rainfall

Rainfall is a fundamental ecological factor. It is one of the main elements of climate that affects the plants on which wildlife feeds, but it is also responsible for the presence and density of certain animals in a given environment. From the Table 2 showing recorded rainfall shows that the wettest months during the three years under study were March with an average rainfall of 185.3 mm, January with 250 mm and March with 187 mm. On the other hand, July and August were the driest months, with an average rainfall of 0 to 0.5 mm.

Table 02: Monthly variations in precipitation (in mm) at the Tizi-Ouzou station, period 2016, 2017 and 2018.

	Jan	Feb	March	Apr	May	June	July	August	Seven	Oct	Nov	Dec
2016	9.9	96.2	185.3	61.8	68.4	5.9	0	0	4.2	16.6	68.4	150.1
2017	250	36	29	37	2	8.8	0.4	0	37.8	35.3	126	138
2018	31	108	187	102.0	33	34	0.5	0	42	182	175	44.0

1.2.3. Humidity atmospheric

It is a state of the climate that corresponds to the amount of water contained in the air. The atmospheric humidity recorded during the 2016-2018 study period shows that the wettest month is December with 85% and 83% respectively, the least humid month is July with humidity levels of 55%, 49% and 59% respectively (Tab. 3).

Table 03: Monthly variations in relative air humidity (in %) at the Tizi station.Ouzou, 2016, 2017 and 2018.

	Jan	Feb	March	Apr	May	June	July	August	Seven	Oct	Nov	Dec
2016	76	73	76	75	71	59	55	59	62	65	72	85
2017	82	75	72	68	62	58	49	56	61	66	76	83
2018	79	80	75	75	80	68	59	60	67	76	77	83

1.2.4. Summary climate

1.2.4.1. Umbro-thermal diagram of Bagnouls and Gaussen

Bagnouls and Gaussen (1953) consider that a dry month is one in which the total rainfall expressed in millimetres is less than or equal to twice the temperature expressed in degrees Celsius. The diagram is designed so that the rainfall scale P expressed in millimetres is equal to twice the average monthly temperature (T) expressed in degrees Celsius, i.e. P = 2T.

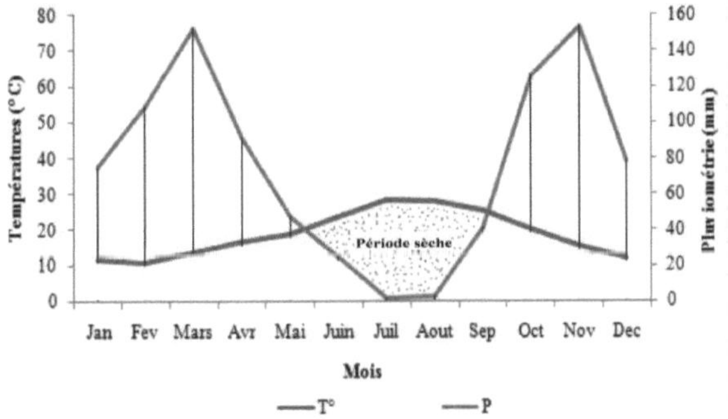

Figure 9: Bagnouls and Gaussen umbro-thermal diagram for the Tizi-Ouzou region (2007-2018).

In the Tizi-Ouzou region, the umbro-thermal diagram (Fig.9) shows that the drought is particularly intense in summer, starting in May and lasting until the end of September, when it lasts for 5 months. As for the wet period, it extends from the beginning of January to the end of May, when it is interrupted by the dry period, and resumes again towards the end of September until the end of December.

1.2.4.2. EMBERGER Climagram applied to the study region

In order to situate our study region in the Emberger climagram for Algeria, the annual averages (minimum and maximum) of temperature and rainfall are essential for calculating the Emberger quotient simplified by Stewart, which is written as follows:

$$Q_2 = 3.43 \times P/M\text{-}m$$

P: average annual precipitation (mm).

M: Average temperature of all the maxima in the warmest month. m: Average temperature of all the minima in the coldest month.

The Emberger rainfall quotient (Q_2) calculated for Tizi-Ouzou for a period of 10 years will therefore be : **Q2 = 3.43 x 868.26 / 36.2 - 7 = 102**

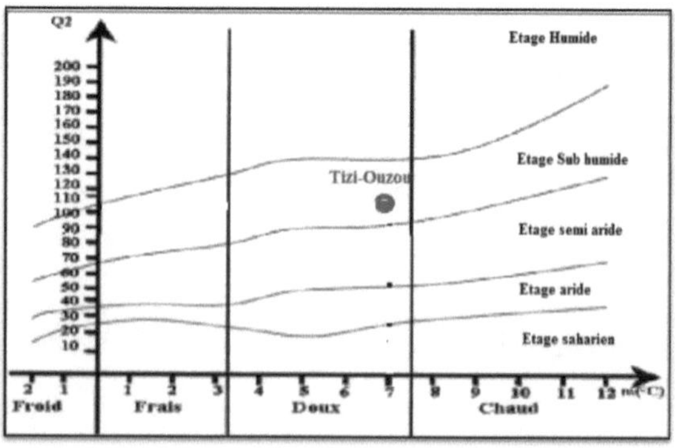

Figure 10: Emberger climagram for the Tizi-Ouzou region (2008 to 2018).

The two values Q2 and m place our study region in the subhumid bioclimatic stage with mild winters (Fig.10). The distribution of regions on the Emberger climagram is influenced by topography and climatic factors.

1.3. Vegetation

Plants are the Apoids' exclusive food source, and vegetation also provides a habitat for bees nesting in the wood or stems of plants.The land-use map shows the importance of hardy arboriculture, with olive groves largely occupying the slopes of the central massif (the Kabyle massif) on slopes often exceeding 25%. The Sébaou and Draâ El Mizan plains are ploughed and cultivated almost all year round. Vineyards and arboriculture account for 52% of the total surface area occupied by the plains.Forests dominated by oak and cedar come second, with 34%. This distribution highlights the agro-sylvo-pastoral nature of the region. Vegetation is generally very dense in Kabylia (Asla, 2002).

1.4. hydrographic network

Water resources are of particular interest to flora and fauna. Tizi-Ouzou's hydrographic network comprises two major catchment areas: the Oued Sébaou basin and the coastal basin. The lower part of the basin has a high water table, due to its essentially impermeable subsoil. In the mountains, the infinite number of These water sources are supplied by the high rainfall and melting snow of the Djurdjura massif. This feeds the Oued Sébaou, which acts as the main collector of the hydrographic network. The Taksebt dam, which holds 175 million m^3 of water, was built in 2001 at Oued-Aissi to improve the supply of drinking water to the populations of Tizi-Ouzou, Boumerdès and Algiers.

1.5. Nature of the soil

The type of soil has a major influence on the type of vegetation. Not only does the soil provide the mineral elements needed for plant growth, which in turn becomes a source of food for flower-feeding insects, particularly bees, as it has a considerable influence on the nectar intensity of flowers, it also provides a home for wild terricultural bees. From a lithological point of view, flysch and schist, formations that are moderately resistant to erosion, predominate in the coastal mountains and the Kabyle massif. Marls and clays, which are not very resistant to erosion, occupy a much smaller area compared with the former; they are located in the two depressions: Sébaou (which includes Tizi-Ouzou) and Draâ El Mizan (Asla, 2002).

2. Station study

The investigations were carried out during 2016, 2017 and 2018 at the Institut Technique Moyen Agricole Spécialisé (I.T.M.A.S) in Boukhalfa, located 8 km from the town of Tizi-Ouzou and bounded to the north by the Oued Sébaou. The geographical coordinates of this institute correspond to 36°44\`46\`\`N latitude, 04°01\`18\`\`E longitude and 89 m altitude. The plot measures 160 m^2 (40 m long by 4 m wide) and is located within a citrus orchard (orange, mandarin and lemon trees) and the institute's experimental plots. The plant species recorded include: Oxalis pes-caprae L. (Oxalidaceae), Sinapis arvensis L. (Brassicaceae), Urospermum dalechampii L. and Centaurea pullata L. (Asteraceae), Convolvulus tricolor L. and Convolvulus arvensis L. (Convolvulacea) (Fig.11).

Figure 11. Photo of the I.T.M.A.S plot (Boukhalfa) (Original photograph, 2016).

2.1. Plants studied

Observations were made on broad beans (Vicia faba L.), onions (Allium cepa L.), lettuce (Lactuca sativa L.), courgettes (Cucurbita pepo L.) and watermelons (Citrullus lanatus .) in 2016 and 2017, and on chickpeas (Cicer arietinum L.) in 2017. Finally, the study on orange trees (Citrus sinensis L.) was carried out in 2016, 2017 and 2018.

Table 4 gives the variety, sowing and flowering dates and planting system for each of the plants studied.

Table 4. Varieties, sowing dates and density of the different plants studied.

Plants	Varieties	Sowing date	Start date of flowering	density
Beans Vicia faba L	major	15 December 2015 06 December 2016	22 February 2016 15 February 2017	25 plants/m2
Chickpeas Cicer arietinum L.	Kabuli	02 January 2017	23 April 2017	25 plants/m2
Lettuce Lactuca sativa L.	Wafia	17 March 2016 24 March 2017	14 June 2016 23 June 2017	20 plants/m2
Onion Allium cepa L.	Dark red D'Amposta	21 December 2015 27 December 2016	12 May 2016 25 May 2017	20 plants/m2
Courgette Cucurbita pepo L.	Green Algiers	08 March 2016 17 March 2017	12 May 2016 18 May 2017	2 plants/m2
Watermelon Citrullus lanatus Thunb.	Charleston gray	02 June 2016 25 May 2017	04 May 2016 02 May 2017	2 plants/m2
Orange tree Citrus sinensis L.	Varietal mix	/	02 March 2016 05 April 2017 27 March 2018	

3. Methods for studying apoidal fauna and its impact on grain yields

3.1. Inventory and density of pollinating fauna

To carry out the inventory of foraging insects on the various plants studied, we opted for two different methods, depending on whether the plants were market garden plants or fruit trees (citrus).

a. Plants

The quadrat method is used for observation, counting pollinators and floral density (Sonnet and Jacob-Remacle 1987; Abrol 1988). Nine quadrats of 1 m^2 each are delimited by means of fils and stakes in the first two rows from the edge of the crop. The distance between two quadrats is 50 cm (Fig. 12 a). All the quadrats are exposed to the sun and receive the same amount of water.

Every two days, from floraison onwards, the observer walks through the quadrats from 7 a.m. until 4 p.m. (Gmt + 1). In each of them and for 7 to 8 minutes, he counts the foraging insects. A few specimens of each species encountered are captured for identification in the laboratory and the creation of reference material.

At the same time as counting, the density of open fleurs is effected in each of the quadrats. Insect density is thus estimated by relating the number of insects counted to a defined number of 100 fleurs. This method is currently the most frequently used to compare pollinator densities in crops (Pierre et al. 1997, 1999).

b. Orange tree

The citrus orchard covers an area of 0.8 ha. The number of trees in the orchard is 149 (101 orange trees, 42 mandarin trees and 6 lemon trees). Flower phenology is determined by counting the number of flower clusters formed on five branches at each cardinal point (North, South, East, West) of 5 randomly selected trees (De Grandi- Hofman, 1988; 1990 cited by Louadi, 1999a). The number of bunches is multiplied by the average number of flowers per bunch. Flower density was determined every two days during flowering.

The bees were counted using the transect method adapted to fruit trees (Jacob-Remacle, 1989). It involves five 50 m rows, each comprising 10 trees (fig. 12b). Apoids were sampled in the orchard three times a week. Bee density was estimated per hectare.

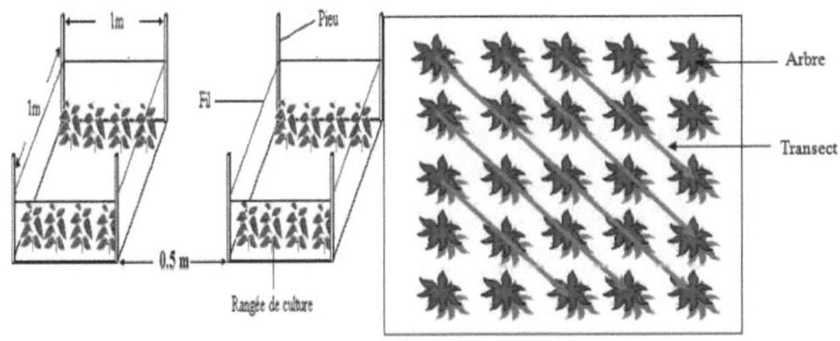

a. Quadrat methodb Transect method

Figure 12: Methods for studying apoid fauna in cultivated environments (Benachour, 2008, Cherair, 2016).

4.3.3. Technique for capturing Apoids

According to Pierre et al (1997), species determination should be carried out on sight during the course of observations specific to the experiment, so as not to modify the structure of the insect population by sampling. For little-known species, this involves capturing a few individuals of the species to be determined, so that they can then be easily identified in the field. Trapping is not recommended for this type of study, for two reasons: firstly, trapping is likely to be selective (some species are less likely to be trapped than others), and secondly, only those individuals that actually visit the plant for pollination should be considered (Pierre et al., 1997).

4.3.3.1. Hunting at sight

According to Meriguet et al (2004), this method is used in sunny weather using a butterfly net to capture large, fast-flying bees such as Xylocopinae, Anthophorinae and Bombinae. The captured insects are placed in a plastic bag containing a cotton pad soaked with of ethanol so that the bees die quickly without being damaged.

4.3.3.2. Direct capture using plastic boxes

This technique is carried out by direct approach using plastic boxes 11cm high and 8cm in diameter, containing a cotton pad soaked in a few drops of ethanol. Several species of bee can be easily captured using this method, which allows for efficient hunting while reducing the risk of breakage and injury, and is particularly useful for harvesting small species.

3.1.2. Identification technique for Apoïdea

Identifying a bee down to its specific rank is not easy. It is done with the aid of a binocular magnifying glass, using various determination keys that allow the different bee genera and species to be separated. The specimens captured during this study were determined mainly using the following keys: Patiny and Terzo (2010), Terzo and Rasmont (2010), Pauly (2014), Rasmont and Terzo (2010).

Determination was also carried out using reference material from the Zoology laboratory at the University of Mons Hainant (Belgium) and the Entomology laboratory a t the Mouloud Mammeri University in Tizi-Ouzou.

Collected specimens are pricked with appropriately sized entomological pins and labelled. Rigorous labelling is essential to avoid any loss of data and information relating to each specimen (place and date, host plant, etc.). Each specimen bears 3 labels (Fig. 13).

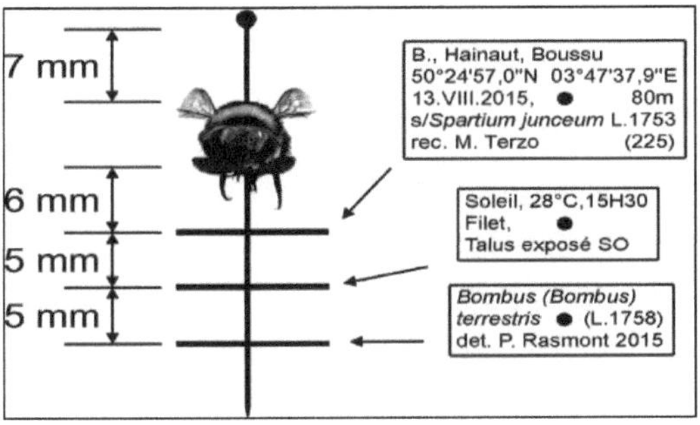

Figure 13: assembly of Apoide specimens (According to the Zoology Laboratory, Mons).

3.2. Foraging behaviour of insects

The foraging behaviour of the most frequent species is observed on the plant from the opening to the closing of the flowers in full bloom. We observed foraging behaviour in order to understand the nature of visits (purpose of visit). We note the feeding purpose of the visits made by the various pollinators; at the same time, we observe whether or not the foraging insect touches the flower stigma. As in other work on pollination, visits with stigma-insect contact will be described as pollinating or positive visits (Jacobe-Remacle, 1989).

To assess pollination efficiency, we measured the number of flowers visited per minute. The average number of flowers visited per minute was then calculated to estimate the speed of the bee species observed.

3.3. Evaluation of the impact of entomophilic pollination on yield grain

In order to study the impact of entomophilic pollination on bean and onion grain yield during floraison in 2016, 18 quadrats of 1m^2 each were set up. Nine of these quadrats are left free for pollinators to access, while the other nine quadrats are caged with tulle (for onion, only the inflorescences are wrapped) afin to prohibit pollinator access (Fig.14). At harvest time, the ripe bean and onion seeds are shelled. For each quadrat, the weight and total number of seeds harvested per plant and per inflorescence are established to compare the yield of seeds from caged quadrats with that from quadrats accessible to bees.

Beans Onions

Figure 14: Photograph of the caged quadrats and the quadrats left free for the bees to access (Original photograph, 2016).

3.4. Statistical analysis of data

The Mann Whitney U test (used for small numbers or when the distribution of the data is not normal) is used to compare the means in the free and caged quadrats for the study of the effect of entomophilic pollination on plant yield, foraging speed and the time spent by bees on the flowers of the two Cucurbitaceae. A one-factor analysis of variance (ANOVA) was used to compare the speed at which honeybees forage on orange flowers at different times during their activity. A normality test (Shapiro-Wilk test) was used to verify the normality of the distribution of the data for the different averages calculated and the different statistical tests carried out. The various statistical analyses were carried out using Statistica version 7 and Statbox version 6.40 software. Graphs were produced using Excel version 2007.

CHAPTER III
RESULTS AND DISCUSSION

At the end of this study, several aspects are developed. Firstly, the flowering of each of the different plants studied is described. In addition, our study enabled us to understand the foraging activity of apoids and to understand the diversity and density of pollinators in each of the crops studied (broad bean, chickpea, onion, courgette, watermelon, lettuce, orange).There are several criteria for assessing the pollination efficiency of a foraging bee. To do this, we estimated the quality or efficiency of foraging by the main bee species that we observed on the flowers of the plants studied. We determined the nature of their floral visits and measured their foraging speed.The pollination efficiency of the bees encountered is also assessed at crop level, highlighting their impact on crop yield. This last aspect concerns only broad beans (Vicia faba) and onions (Allium cepa), and our observations show the effect of entomophilic pollination on the grain yield of these very common and widely cultivated plants.

1. Fava beans (Vicia faba L.)

1.1. Flowering of the plant

To find out whether there is a synchronisation between the temporal evolution of the number of bees and that of the flowers of the foraged plant, in this case V. faba, we proceeded to count the bees and the flowers simultaneously. Several cultivated plant species have short spring flowering periods, including broad beans. During the two years of observation in 2016 and 2017, V. faba flowering began on 22 February and 15 February respectively and lasted 48 and 33 days. A faba bean plant bears between 27 and 58 flowers (33 flowers/plant on average) grouped in inflorescences. We counted an average of 7 inflorescences per plant and 4 flowers per inflorescence. During flowering of V. faba, the flowers first develop at the base of the stems, gradually moving upwards over the following week. Bees colonise the plot as soon as the first faba bean flowers appear. The number of bees gradually increases over time and follows the evolution of the number of V. faba flowers. In fact, the highest number of visits is recorded during the period of full flowering of this plant, reaching its peak on 15 March 2016 and 01 March 2017 with respective abundance peaks of 166 and 187 individuals (Fig. 15 a and b). Towards the end of flowering, bee density gradually decreases as flowering declines.

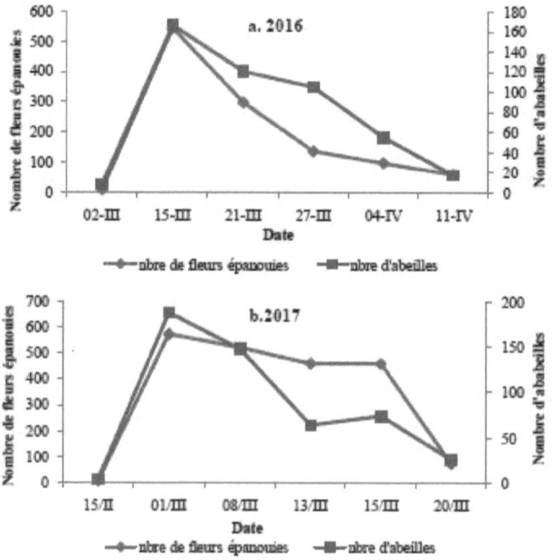

Figure 15: Flowering cycle of V. faba and bee density during the 2016 (a) and 2017 (b) flowerings.

1.2. Diversity and density of pollinators

The study carried out on V. faba during the two flowerings revealed that this legume attracts a pollinating entomofauna composed almost exclusively of apoid Hymenoptera belonging to three families: Apidae, Megachilidae and Andrenidae. Seven species have been recorded, but only three of them were observed during the two flowering periods: Eucera numida Lepeletier 1841, Apis mellifera L. 1758 and Xylocopa violacea L. 1758.
The numbers of visits observed for the two flowering periods studied are given in Table 5. The results show that the most abundant species on the bean flowers is the solitary bee E. numida, which alone represents more than half of the total number of bees (64%) and a density of 8 to 17 individuals per 100 flowers (Tab. 6). It is also the only species to be present on the flowers throughout the flowering period. In second place was the honey bee A. mellifera with 31%. On the other hand, X. violacea and other species are present only occasionally.

Table 5: Number of specimens (Nind) and relative frequencies (%Nind) of apoids recorded during transects.

Family	Species	N	N%
	Eucera numida Lepeletier 1841	494	64
	Apis mellifera L., 1758	243	31
Apidae	Bombus terrestris L. 1758 Xylocopa violacea L. 1758	16 12	2 1,5
	Bombus ruderatus Fabricius, 1775	8	1
Megachilidae	Osmia rufa L. 1758	1	0,13
Andrenidae	Andrena. Sp.	1	0,13
Total	7 species	775	100

Table 6: Average number of pollinators per 100 flowers of V. faba during the 2016 and 2017 flowering seasons

Family	Species	2016	2017
	Eucera numida	8	17
	Apis mellifera	3	4
Apida	Xylocopa violacea	<1	<1
	Bombus terrestris		1
	Bombus ruderatus		<1
Megachilidae	Osmia rufa		<1
Andrenidae	Andrena sp.	<1	

1.3. Foraging activity of apoids

During the two V. faba flowering periods (2016 and 2017), the Apoids begin to appear at 9am and 10am respectively, when the first rays of sunlight appear and the temperature reaches an average of 18°C and the relative humidity approaches 60%. In the hours that follow, we gradually see an increase in the number of foraging bees and the colonisation of the floral area. The evolution of the number of bees (all species taken together) during the different hours of the day (9-15h) is shown in figure 16. Our results show that bee activity changes from hour to hour. The bees seem to concentrate their floral visits between 12pm and 3pm, a period that corresponds to the optimal opening of a large number of bean flowers likely to be foraged. During both flowering periods, our observations recorded a maximum number of foragers at 3pm.

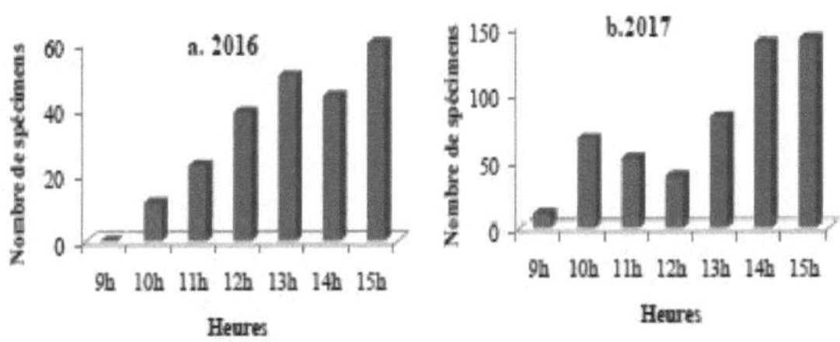

Figure 16: Total number of apoids recorded at different times of day during the 2016 (a) and 2017 (b) blooms.

1.4. Foraging behaviour and efficiency pollinator

The foraging behaviour of the two most frequent species on V. faba flowers, E. numida and A. mellifera, was observed during flowering in 2016 and 2017. We noted that E. numida always practises positive foraging by penetrating inside the flower (Fig.17 a, b). We also noted that 47% of these visits concerned the collection of nectar, 13% for pollen and 40% for the mixed collection of pollen and nectar (Tab. 7).

The honey bee only forages positively when it collects pollen, i.e. on only 28% of its visits. On the other hand, to collect nectar (72% of visits), they always use the holes drilled by the bumblebees at the base of the corolla and the extra-floral nectaries (Fig. 17 c), drawing nectar without touching the stamens, which does not allow the flower to be pollinated. As for the frequency of visits, our results show that the honey bee visits more often than the honey bee (Tab. 8). If we examine the numbers and pollination efficiency of the two flowerings for these two bee species, we see that 88% of the pollination of the broad bean is carried out by E. numida. We can therefore deduce that this species is the most efficient on this plant.

Table 7. Distribution of floral visits (in %) according to the floral product harvested and their pollination efficiency during the 2016 and 2017 flowering seasons
P: pollen; N: nectar; +: pollinating visit; -: non-pollinating visit

	Eucera numida	Apis mellifera
Number of visits observed	494	243
P+	13%	28%
N+	47%	0%
N-	0%	72%
(P+N)	40%	0%
Total pollinating visits	100%	28%

Table 8: Number of visits observed and pollinating visits by the two most abundant species on V. faba flowers (only one visit is counted per insect).

Flowering	Species	Number of specimens observed (N)	N%	Number of pollinating visits (P)	P%
2016	Eucera numida	106	40	106	85
	Apis mellifera	162	60	18	15
2017	Ecera numida	388	83	388	88
	Apis mellifera	81	17	51	12
Total	Eucera numida	494	67	494	88
	Apis mellifera	243	33	69	12

-a- male-b- female-c-

Figure 17: Positive foraging by Eucera numida (a, b) and negative foraging by Apis mellifera. (c) on bean flowers (original photograph, 2016).

1.5. Efficiency pollinator

We assessed the pollination efficiency of two species, A. mellifera and E. numida. To distinguish male and female E. numida in the field, it is sufficient to observe the length of their antennae, which are much longer in the male (Fig. 17 a and b). Pollination efficiency is also assessed by the speed at which the bees forage. The distribution of bee species according to the number of flowers visited as a function of time (number of flowers per minute) (Tab. 9) shows that the most frequent visits are those of the male E. numida (figure 17a), which forages flowers at an average rate of 9 flowers per minute (fls/mn). As for the females of this same species (figure 17 b), they seem to be slower than the males and as fast as the honey bee (6 fls/mn), visiting the flowers at an average rate of 7 flowers/mn. The results of the Mann Whitney Test ($p<0.05$) (Tab. 9) show that during flowering in 2017, male E. numida was significantly faster than female.

Table 9: Average number of V. faba flowers visited per minute by the main species of bee during the 2017 flowering season.
(±: standard error; N = number of specimens observed). (Mann Whitney test (U); p<0, 05)

Species	N	speed	U	P
Eucera numida (female)	32	7 ± 1.36	109	0
Eucera numida (male)	32	9 ± 1.4		
Apis mellifera	32	6 ± 1.13		

1.6. Searching for food bees

The activity of E. numida on bean flowers begins at 10 a.m., with the majority of floral visits devoted to nectar collection and mixed collection of the two products (Fig. 18 a). The highest proportions of nectar collection were observed at 10 and 11 o'clock in the morning. As for mixed harvesting, the highest proportions were recorded at 1pm. Pollen was collected mainly at midday and 2pm. It was during the afternoon, from 1 to 3pm, that the highest numbers of visits of all types were recorded. Conversely, the honey bee's floral visits are always mainly concerned with collecting pollen (Fig. 18 b), at all times of the day. As in the case of the Eucera, the greatest number of sightings of the honey bee occurred between 1pm and 3pm. In addition, we found that this bee is not very loyal to the bean and visits many other wild plants. This explains the low densities of foragers recorded on fava beans during the 2016 and 2017 flowering seasons.

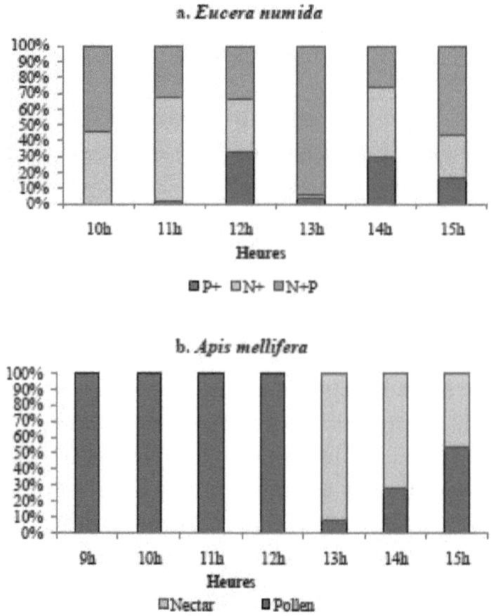

Figure 18: Percentages of floral products harvested by E. numida (a) and A. mellifera (b) from V. faba at different times of the day.

1.7. Evaluation of the effect of entomophilic pollination on bean yields

The study of the effect of insect pollination on plant production during the 2016 flowering season shows that the presence of pollinating insects greatly improves the yield of the broad bean crop.

1.7.1 Effect of entomophilic pollination on pod yield

The length and average number of green pods from free quadrats were significantly greater than those from covered quadrats (Tab. 10).

Table 10: Average length and number of green pods during the fruiting phase V. faba during flowering in 2016. (U = Mann whitney test).

	Free quadrats	Caged quadrats	P	U
Pod length (cm) n=number of pods	(n=130) 16,41± 3.4	(n=130) 13.03 ± 2.86	0	3817.5
Number of pods/plant n=number of plants	(n=30) 10.46 ± 4.3	(n=30) 5.23 ±3.4	0	151

1.7.2. Effect of entomophilic pollination on seed yield

The average seed weight of the free quadrats was three times higher than that of the caged quadrats. Average harvest weights per plant were also higher on plants accessible to pollinators than on those caged. The average seed weight calculated from 30 samples of 5 seeds each drawn randomly from the harvest was also significantly greater in the free quadrats than in the caged quadrats (Tab.11).

Table 11: Average grain yield of V. faba in free (QL) and caged (QE) quadrats during flowering in 2016.(U = Mann whitney test; t = Student's t test).

	Free quadrats	Caged quadrats		
Average harvest/quadrat weight (g) (n: number of quadrats)	(n=9) 544.44 ± 113.52	(n=9) 197.22 ± 63.03	t=8.05	P=0
Average seed weight (g) (n: sample of 5 seeds)	(n=30) 10.04 ± 1.73	(n=30) 5.68±1.36	t=10.84	P=0
Average seed weight/plant (g)(n: number of plants)	(n=30) 50.68 ± 25.38	(n=30) 16.19 ± 11.04	U=66	P=0

The average number of pods and seeds per plant was also higher in free plants than in caged plants (Tab.12). As for the number of aborted seeds, it was much higher in the case of foraged plants than in the case of caged plants in the absence of bees.

Table 12: Average Vicia faba fertility indices during flowering in 2016.

	Free quadrats	Caged quadrats
Average number of pods / plant	10.5	5.23
Average number of seeds/plant	28	12.43
Total number of aborted seeds	921	664

1.8. Discussion

A study of the density and activity of Apoids during the flowering of the faba bean has shown that their numbers depend on the floral density of the plant. Their numbers increase gradually over time and follow changes in the number of V. faba flowers. Our study therefore reveals a major influence of the number of flowers on the density of foraging bees. These results corroborate those of Aouar (2009), who noted that fluctuations in the density of foraging bees and the phenology of the faba bean occur synchronously. Observations carried out during the 2016 and 2017 flowering seasons reveal that the plant's pollinating entomofauna in the Tizi-Ouzou region is composed mainly of apoïdes Hymenoptera of the Apidae family.Eucera numida was the most abundant species on the flowers during both flowerings. On average, it accounted for 64% of the visits observed. This situation is similar to that observed in Constantine (Benachour et al., 2007) and in southern Spain (Pierre et al. 1999), where the same Eucera constituted 70% and 89.4% respectively of the insect pollinators observed on the flowers of cultivated fava beans. The honey bee comes second in terms of abundance. These resultsare consistent with those observed in Tizi-Ouzou (Aouar, 2009) and Constantine (Aouar, 2009).(Benachour et al., 2007). Tasei (1976) reported that the honey bee accounts for nearly 80% of the apoid fauna foraging on faba bean. In Finland, the honey bee is the most abundant species on faba bean with a visit rate of 41% (Kyllönen, 2018). Xylocopa violacea, Bombus terrestris and Bombus ruderatus are very poorly represented. The latter are considered to be good pollinators of the plant (Stoddard & Bond 1987; Pierre et al. 1997, 1999). This low abundance may be explained by the fact that other plants whose flowering period coincides with that of the bean are also frequented and preferred.

On the other hand, with the exception of a single species of Andrenidae, all the bees inventoried are members of the Apidae family, characterised by a long glosse, giving them an increased ability to collect nectar (Pesson and Louveaux, 1984). Collecting nectar is all the more difficult when the base of the flower is deep, as in the case of broad bean flowers. In Egypt, the bees that visit the faba bean are both long-tongued and short-tongued (Shebl and Farag, 2015).

A study of the foraging behaviour of E. numida and A. mellifera shows that the Eucera always forages positively on the flowers, whatever the product harvested, and all its visits can therefore be fertilising. In honeybees, the proportion of positive foraging is on average 28%, which corresponds to the collection of pollen. According to Somerville (2002), broad beans are mainly pollinated by pollen-collecting bees. As for nectar collection, this always corresponds to non-fertilising foraging, as the bee takes advantage of the extra-floral nectaries and the holes pierced at the base of the corollas by B. terrestris. This foraging behaviour has also been observed by several authors (Stoddard & Bond 1987; Free 1993; Koltowski 1996; Pierre et al. 1997; Pierre et al. 1999, Benachour et al. 2007; Aouar, 2008; Marzinzig 2018 and Kyllönen, 2018).Concerning foraging speed, it appears that male E. numida has the highest foraging speed compared to female E. numida and the honey bee. Our results corroborate those of Aouar et al (2008) who observed that the number of flowers visited by Eucera is fairly high. Benachour et al (2007) estimated the speed of E. numida at an average value of 10.4 flowers per minute. This value is very close to that recorded for the same species, which forages at a rate of 9 flowers per minute. According to Tasei (1990), the speed of the honey bee is 6 to 8 flowers/minute when foraging for nectar, and 11 to 16 flowers/minute when foraging for pollen. These results are similar to our own, which estimated the average foraging speed of the honey bee at 6 flowers/minute. The daily activity of bees on flowers during the bean flowering period depends on the production of pollen and nectar, the quantities of which vary throughout the day (Pierre et al. 1996; Pouvreau 2004). Our results show that the bees' activity changes from one hour to the next, and they seem to concentrate their floral visits between 12 and 3pm, a period that corresponds to the optimal opening of a large number of bean flowers likely to be foraged. These results are consistent with those of Benachour et al (2007), Aouar et al (2008) and Shebl and Farag (2015). Our studies in 2016 and 2017 show that the major part of floral visits by Eucera are devoted to nectar collection, the largest proportions of which are observed at 10 am and 11 am. Pollen is collected mainly at midday and 2pm. Our observations are similar to those of Benachour et al (2007), who noted that the proportion of visits devoted to nectar collection is greatest from 10 to 11 am and from 3 to 4 pm. According to Somerville (2002), nectar is mainly secreted early in the morning and late in the evening. Conversely, honey bees visit flowers mainly to collect pollen at all times of the day, but this species is not very faithful to the bean and visits many other wild plants. This explains the low densities of foragers recorded on fava beans during the

2016 and 2017 flowering seasons. The presence of pollinating insects helps to improve the production of V.faba. The grain yield and number of pods obtained by cross-pollination are higher than those obtained by autogamy. In fact, our observations show that the weight of seeds from free quadrats is three times greater than those from caged quadrats. Our results corroborate those obtained by several authors, including Mesquida et al. (1990), Varis & Brax (1990), Koltowski (1996), Musallam et al. (2004), Benachour et al. (2007), Aouar et al. (2008), Gasim and Abdelmula (2018) and Marzinzig et al. (2018).
With regard to pod size, we found that pods were longer in the free quadrats than in the covered quadrats. These results are in line with those of Aouar et al (2008) who found that pods from open plots measured an average of 24.8 cm compared with 21.1 cm for those from caged plots. According to Barbier, quoted by Aouar (2009), cross-pollinated or entomophilous pods are longer and contain more seeds than self-pollinated pods.
According to Tamas et al (1979) and Stephenson (1980), quoted by Benachour (2008), the high number of aborted seeds in open plots can be explained by the fact that plants in open plots, whose flowers are fully pollinated, cannot ensure that all the seeds ripen. Pods from the first flowers pollinated flowers reach maturity, while those from the last pollinated flowers abort at the young pod stage.

2. Chickpea (Cicer arietinum L.)

2.1. Flowering of the plant

Flowering of this legume began on 23 April 2017 and lasted 25 days. During flowering of C. arietinum, the flowers first develop at the base of the stems, with development continuing gradually towards the top over the following week.
Bees can be seen as soon as the first chickpea flowers appear. Their numbers increase progressively over time and follow the evolution of the number of flowers of this legume. The highest rates of visits are recorded during the period of full flowering of this plant between 03 May and 10 May (2017) (Fig.19). Towards the end of flowering, the density of bees gradually decreases as flowering declines.

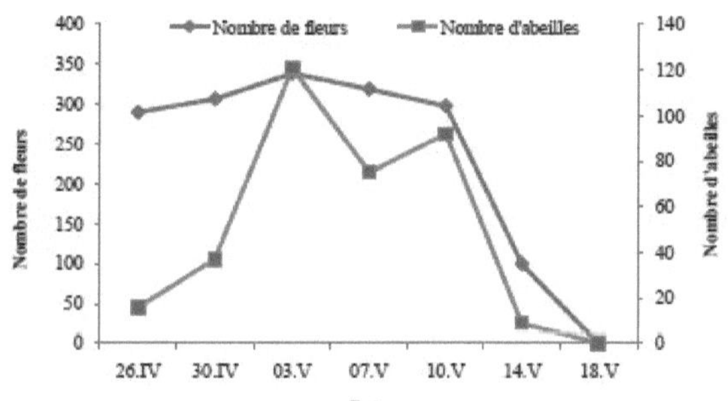

Figure 19: Flowering cycle of Cicer arietinum and bee density during flowering.of 2017.

2.2. Diversity and density of pollinators

The study carried out on C. arietinum L. shows that this legume attracts a pollinating entomofauna composed almost exclusively of apoid Hymenoptera belonging to three families: Apidae, Megachilidae and Andrenidae. Eight species have been recorded, but only two are dominant: Eucera decolorata Gribodo, 1924 and Eucera eucnemidea Dours, 1873 (Tab.13), which together account for more than half of the total population. of the total number (72%). These two species have an average density of 9.24 and 9 individuals per 100 flowers respectively (Tab. 14). In third and fourth place come the honey bee Apis mellifera and Lithurgus chrysurus respectively (3.1 and 2.8 individuals). Anthophora quadricolor and other species, on the other hand, occur only occasionally.

Table 13: Number of specimens (Nind) and relative frequencies (%Nind) of apoids recorded during transects

Family	Species	N	N%
	E. decolorata Gribodo, 1924	126	37.61
Apidae	E. eucnemidea Dours, 1873 A. mellifera L. 1758	114 41	34.03 12.24
	Anthophora quadricolor Erichson, 1840	7	2.09
	Lithurgus chrysurus Fonscolombe, 1834	43	12.84
Megachilidae	Megachile sp.	2	0.60
	Anthidium sp.	1	0.30
Andrenidae	Andrena sp.	1	0.30
Total	8 species	335	100

Table 14: Average number of pollinators per 100 flowers of Cicer arietinum during the 2017 flowering season.

Family	Species	Density
	E. decolorata	9.24
Apidae	E. eucnemidea A. mellifera	9 3.1
	Anthophora quadricolor	<1
	Lithurgus chrysurus	2.8
Megachilidae	Megachile sp.	<1
	Anthidium sp.	<1
Andrenidae	Andrena sp.	<1

2.3. Foraging activity of apoids

Daily activity and foraging behaviour of E. decolorata, E. eucnemidea, A. mellifera and L. chrysurus were observed during chickpea flowering. All four species began foraging on the flowers at 9am, intensifying their activity at 10am and 11am, before dropping off at 12pm. Bee activity resumed at 1pm, then fluctuated over the following hours (Fig. 20).

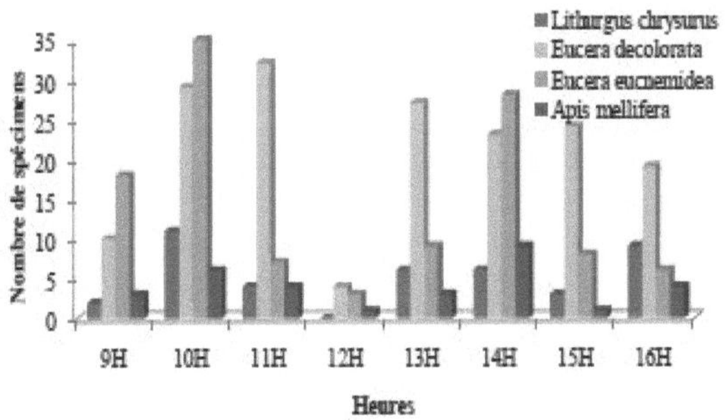

Figure 20: Daily bee activity on C. arietinum flowers during the 2017 flowering season.

2.4. Foraging behaviour and efficiency pollinator

A study of the behaviour of bees observed on the plant did not reveal the collection of bees. pollen by these insects. All visits are devoted to collecting nectar.As for the study of pollination efficiency, we observed the speed of foraging by E. decolorata and E. eucnemidea. To do this, we measured the number of flowers visited in one minute by around thirty individuals of each of these bee species. The results, subjected to the Mann Whitney Test ($p<0.05$), show that during the 2017 flowering period, E. decolorata is significantly faster (13.9 ± 1.81 flowers/minute) than E. eucnemidea (8.4 ± 1 flowers/minute) (Tab.15).

Table 15: Foraging speed of E. decolorata and E. eucnemidea on Cicer arietinum flowers during flowering in 2017(Mann Whitney test (U); $p<0.05$; ±: standard error;).

Species	Number of specimens observed	Speed (flowers/min)	U	P
Eucera decolorata	30	13.9 ± 1.81	5	0
Eucera eucnemidea	30	8.4 ± 1		

2.5. Discussion

The study carried out on C. arietinum during the 2017 flowering season showed that the plant is visited by a pollinating entomofauna composed almost exclusively of apoid Hymenoptera belonging to three families: Apidae, Megachilidae and Andrenidae. Floral visits by E. decolorata, E. eucnemidea, A. mellifera and L. chrysurus are abundant compared with those of other bees. Megachile sp., Anthidium sp. and Andrena sp. only visited occasionally. Our The results corroborate those of Louadi (1999) in Constantine, where he observed the first two families of apoids on chickpea. These two families (Apidae, Megachilidae) are characterised by a long tongue, which allows them to frequent these flowers, which are characterised by a papilionaceous corolla. Eucera from the Apidae family are the most abundant bees on chickpea flowers. They make up more than 71% of all visits observed on this plant. This situation is similar to that noted in Constantine by Louadi (1999) where E. eucnemidea is the most abundant species on this legume. However, in India, Abdul Latif et al. (2019) noted two species of the Apidae family, Apis dorsata and Apis florea, which marked the highest abundance on chickpea flowers.A study of the daily activity of the bees studied on the flowers of C. arietinum revealed intense activity during the morning phase at 10 and 11 am. These results concur with those of Louadi (1999) who noted a maximum activity of E. eucnemidea at 10 am. Observation of the foraging behaviour of the four most abundant species d i d not reveal any pollen collection by these bees. According to Louadi (1999), the difficulty of accessing chickpea anthers is due to the flower's morphology, which has a closed standard and two closed carinae. As a result, only nectar is collected by the bees. These results confirm the studies carried out in Punjab (India) by Batra (1979) (cited by Louadi, 1999). These revealed that bees visit chickpeas but do not pollinate them. As for the study of pollination efficiency, our results showed that E. decolorata is more efficient than E. eucnemidea. In India, on the other hand, A. florea was the fastest pollinator (Abdul Latif et al., 2019).In conclusion, although chickpea is known to be self-pollinating, several studies show that cross-pollination can occur. Hawtin (1978) estimated that insect pollination is only 2%. Through our observations, eight species of bee were recorded on the flowers of this plant, of which E. decolorata and E. eucnemidea recorded the highest abundances and foraging speeds. The presence of these insects on open-pollinated flowers has resulted in cross-pollination, which will ultimately lead to higher crop yields.

3. Onion (Allium cepa L.)

3.1. Flowering of the plant

The florescences started on 12 May 2016 and 25 May 2017 and were spread over 42 and 24 days, respectively (Fig.21 a and b). An onion plant bears between 1 and 4 inflorescences. The floraison of the plant was significantly reduced in 2017 compared to flowering in 2016, due to a drier winter and a very hot summer characterised by scorching temperatures in 2017.
Colonisation of the plot by bees is noted as soon as the first A. cepa flowers appear. Their numbers increase gradually over time, as the number of A. cepa flowers increases. The highest rates of visits were recorded during the period of full flowering of the onion on 26 May 2016 and 01 June 2017. Towards the end of the flowering period, the number of bees gradually decreases as the flowering declines.

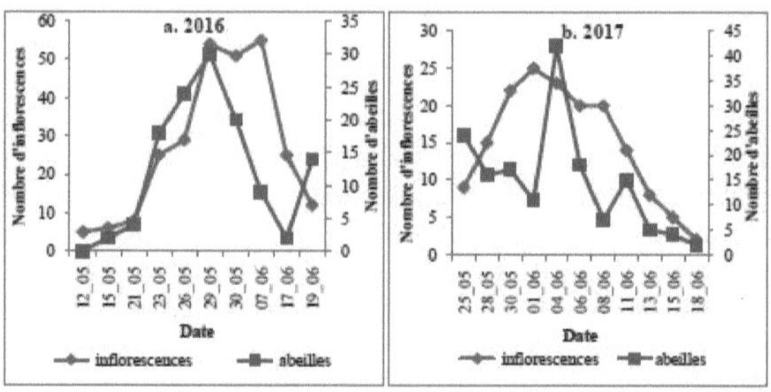

Figure 21: Allium cepa flowering cycle and the number of bees recorded during flowering in 2016 (a) and 2017 (b).

3.2. Diversity and density of pollinators

Observations made during the two floraison periods show that the insects that forage on the fleurs of A. cepa are all apoïdes Hymenoptera belonging to two families: Apidae and Halictidae. Seven species are recorded but only four of them are observed during the two floraisons, these are Apis mellifera, Lasioglossum malachurum, Lasioglossum pauxillum and Halictus gemmeus. In addition, the first three species are relatively abundant. A. mellifera is also the only one to be present on the fleurs throughout the floraison period. The presence of Bombus terrestris and the other species is only episodic. The numbers of visits observed for these three species during the two successive floraisons studied are provided in Table 16.

Table 16: Number of specimens (N) and relative frequency of Apoidea identified on Allium cepa during the 2016 and 2017 blooms.

Family	Species	2016		2017	
		N	N%	N	N%
Halictidae	Halictus gemmeus	2	0.97	2	1.28
	Lasioglossum malachurum	59	28.64	37	23.72
	Lasioglossum pauxillum	47	22.82	46	29.49
	Lasioglossum sp.	1	0.49		
	Sphecodes sp.	1	0.49		
Apidae	Apis mellifera	95	46.12	71	45.51
	Bombus terrestris	1	0.49		
Total	7 species	206	100	156	100

Estimation of the density of bee species found on onion flowers shows that among the Halictidae, L. malachurum and L. pauxillum are the most abundant on fleurs during the two years of observation with an average number of 50 and 43 individuals per 100 flowers for the 1ère species and 42 and 62 individuals for the second (Tab.17). Visits by H. gemmeus were less numerous. Among the Apidae, it is A. mellifera that is frequent on the fleurs, with a density of 37 to 44 individuals depending on the observation period.

Table 17: Average number of pollinators per 100 Allium cepa flowers during the 2016 and 2017 flowering seasons

family	species	2016	2017
	Halictus gemmeus Dours 1872	4	4
Halictidae	Lasioglossum malachurum Kirby	50	43
	1802 Lasioglossum pauxillum	42	62
	Schenck 1853 Lasioglossum sp.	<1	
	Sphecodes sp.	<1	
Apidae	Apis mellifera L. 1758	37	44
	Bombus terrestris L. 1758	<1	

3.3. Foraging activity of apoids

During the 2016 flowering of A. cepa, the Apoids begin to appear from 8am. Over the following hours, there was a gradual increase in the number of foraging bees and colonisation of the flowering area. The evolution of the number of bees (all species) during the different hours of the day (8-14h) is shown in figure 22 a. Our results show that bee activity changes from hour to hour. They seem to concentrate their floral visits between 10 and 11 o'clock, a period which corresponds to the optimal opening of a large number of onion flowers likely to be foraged. During the 2017 flowering period, observations were only made in the morning because of the high temperatures recorded. The bees started flying at 7 am and recorded a maximum number of foragers at 8 am (Fig.22b).

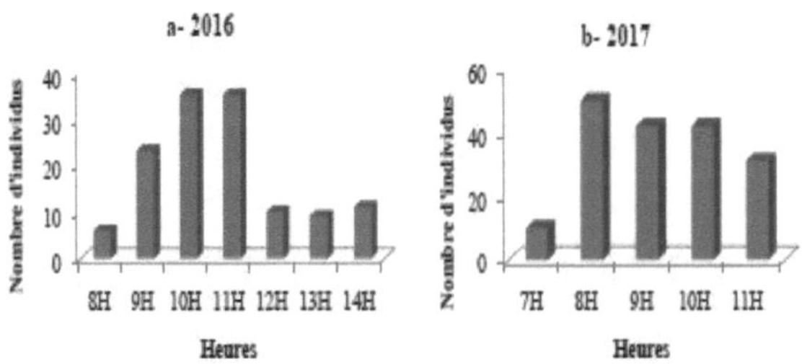

Figure 22: Total number of apoids recorded at different times of day during the 2016 (a) and 2017 (b) flowering seasons.

3.4. Foraging behaviour and efficiency pollinator

The floral morphology of plants influences the foraging behaviour of Apoids: the relative position of stigmas and anthers is an important factor in pollination. Although the onion stigma and anthers are almost at the same level, the latter release pollen before the stigma is receptive. In this study, only the foraging behaviour of the three most common species on fleurs, namely, Apis mellifera, L. malachurum and L. pauxillum is observed. We noticed that all these bees made pollinating visits. In fact, they insert their head into the heart of the flower, thus making a frontal visit (with contact of the stigmas). During the floraison of 2017, 85% of visits to L. malachurum were for mixed harvesting and around 15% for pollen harvesting. As for L. pauxillum, 19% of these visits are devoted to pollen, 8% to nectar and 73% to mixed harvesting (Tab.18). These two species always carry out positive foraging on fleurs. They penetrate inside the flower to collect either pollen or nectar, or both, which triggers the staminal column that causes the release of pollen. As for the honey bee, all its visits are recorded for nectar collection. To collect nectar, the bee rests on the flower while activating the staminal column. Visits by all three species can therefore be fertile.

Table 18. Distribution of visits (in %) of A. mellifera, L. malachurum and L. pauxillum on Allium cepa according to the floral product harvested and effipollination efficiency of these visits during floraison in 2017. (P: pollen; N: nectar; +: pollinating visit; -: non-pollinating visit)

	A. mellifera	L. malachurum	L. pauxillum
Number of visits observed	79	33	64
P+	0	15%	19%
N+	100%	0	8%
N-	0	0	0
(P+N)+	0	85%	73%
Total pollinating visits	100%	100%	100%

If we consider the number of flowers visited per minute, the most frequent visits were recorded for the honey bee, which foraged flowers with an average rate of 13 and 12 flowers per minute (fls/min) in 2016 and 2017. L. malachurum and L. pauxillum came second and third (Tab.19).

Table 19: Foraging speed of A. mellifera, L. malachurum and L. pauxillum on Allium cepa during the two floraison periods (N = number of specimens observed; ±: standard deviation).

Flowering	Species	N	Average number of flowers visited per minute
2016	Apis mellifera	30	13± 3.21
	Lasioglossum malachurum	23	10± 1.6
	Lasioglossum pauxillum	24	9± 1.7
2017	Apis mellifera	30	12± 3.68
	Lasioglossum malachurum	17	11± 1.4
	Lasioglossum pauxillum	21	9± 0.9

3.5. Searching for food bees

Observations carried out during 2017 on flower visits were only carried out during the morning hours because of the high temperatures recorded during the onion flowering period.For species in the genus Lasioglossum, the majority of floral visits during the 2017 floraison, involved mixed nectar and pollen collection (Fig.23, a-b). The greatest proportions of collection are observed from 7am to 9am in the morning. As for pollen, L. pauxillum begins to collect it from 8 am onwards, with a marked increase at 10 and 11 am. L. malachurum starts collecting pollen at 10am, and this collection also peaks at 11am. Nectar collection, on the other hand, is only observed in L. pauxillum, which only harvests it at 8am. Conversely, the floral visits of the honey bee exclusively for nectar collection (Fig.23 c) during all hours of the day.observation.

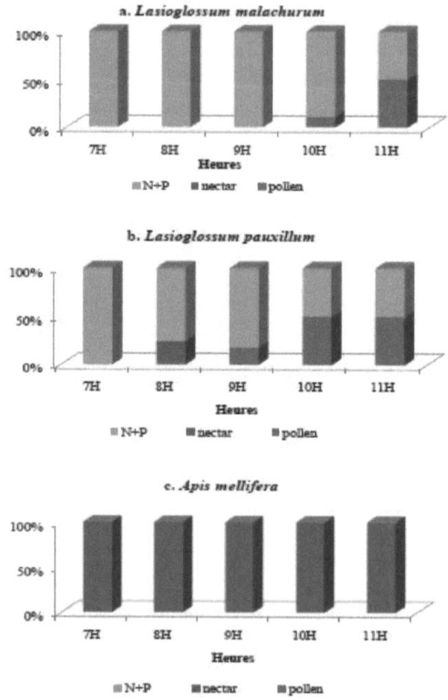

Figure 23: Percentages of floral products harvested by L. malachurum (a), L. pauxillum (b) and A. mellifera (c) on Allium cepa at diff érent times of the morning (floraison 2017).

3.6. Effect of entomophilic pollination on yield grain

Our results showed that the presence of foraging bees considerably improved the yield of the onion crop, recording an economic advantage of around 98% for the average seed weight per quadrat and per plant. The average harvest weight per quadrat and per plant was significantly higher on plants accessible to pollinators than on those caged (Tab.20). The average number of seeds and the number of seed capsules per plant were also significantly higher in the free plots. In addition, the seeds from the free plots appeared larger in size and smoother and more uniform in appearance (Fig. 24).

Table 20: Average grain yield of Allium cepa in the free (QL) and caged (EQ) during the 2016 floraison. (U = Mann Whitney test: $p < 0, 05$).

	QL	EQ	U
Average weight of grain/quadrat (gr)	4.14 ± 3.5	0.09 ± 0.05	0
Average weight of seeds/plant (gr).	1.3 ± 1.4	0.03 ± 0.04	0
Average number of seed capsules/plant	186 ± 203	4 ± 6	0
Average number of seeds/plant	557 ± 611	12 ± 18	0

-a--b-

Figure 24: Onion seeds from free plants (a) and caged plants (b).

3.7. Discussion

Observations made on the onion Allium cepa during the 2016 and 2017 floraisons showed that the fleurs of this Liliaceae attract many visitors, most of which are apoid Hymenoptera from the families Halictidae and Apidae. The honey bee A. mellifera, L. malachurum and L. pauxillum are the most abundant species on the fleurs during the study periods. These results corroborate those of several authors who have studied the pollination of onion crops. In the USA, Nye et al (1973) observed three families of Apoids (Apidae, Halictidae and Andrenidae). In Pakistan, Asif et al (2008) counted only one family, the Apidae, and in Cameroon, Georges et al (2014) counted 3 families, namely the Apidae, Megachilidae and Halictidae.In the present study, the honey bee is the most abundant species, encompassing on average 46% of the visits observed. This situation is similar to that observed by Georges et al. in 2014 in the far north of Cameroon, where the same species constitutes 45% of the insect pollinators of onion flers. L. pauxillum and L. malachurum come second and third respectively in terms of abundance. The other Apoids, notably B. terrestris, H. gemmeus and Sphecodes sp, are very poorly represented. This low density may be due to the attractiveness of other flowering plants in the vicinity of the onion crop. Indeed, in a recent study (Korichi, 2015) that we carried out on the floral choices of bees on the same site, we found that these species were more attracted to spontaneous plants, in particular Echium australe (Boraginaceae), Centaurium umbellatum (Gentianaceae), Hysoseris radiata and Echinops spinosus (Asteraceae). According to Franklin (1970), the lack of attractiveness of onions to bees can lead them to neglect the crop, particularly if another very attractive crop is in flower.

Furthermore, observation of the foraging behaviour of Apoidea shows that all the species encountered effect positive foraging whatever the product harvested, so all their visits can be fertile and beneficial. In this regard, Jones (1937) noted that onion pollination is mainly carried out by insects that visit the nectaries at the base of the three inner stamens. To extract nectar, the apoids must touch the stamens, which triggers the anthers to release pollen. The daily activity of bees on fleurs depends on the production of either pollen (Stone et al. 1998) or nectar (Pierre et al., 1996; Suzo et al., 2001; Pouvreau, 2004) during the day. Our observations in 2017 show that all A. mellifera specimens collect nectar exclusively throughout the flowering period. These results are similar to those of Benedek and Gaal (1972) who found that, on onion inflorescences, bees collect nectar almost exclusively; the number of pollen collectors seen each time during the observations represents less than 1% of the total number of bees. Georges et al (2014) found that 77% of honey bee visits are devoted to nectar collection. On the other hand, according to Jean-Marie (1991), some honey bees forage only for nectar and others for both nectar and pollen. Furthermore, Collison (1973) reports that many pollen-collecting bees later switch to nectar collection. Conversely, according to our study, L. malachurum and L. pauxillum search for pollen and nectar and mainly collect both products at the same time during all the hours of observation. Malachurum spends the first three hours collecting both products, and collects pollen from 10 to 11 am. L. pauxillum, on the other hand, starts collecting pollen at 8 a.m. and only collects nectar at that time of day.

As far as Allium cepa yields are concerned, we found that the presence of insect pollinators makes a major contribution to improving production. The grain yield obtained by cross-pollination is higher than that obtained by autogamy. Concordant results, comparing free and caged quadrats, were obtained by Benedek and Gaal (1972), Zdzislaw et al. (2004), Asif et al. (2008), Mortaza et al. (2013) George et al. (2014) and Aniruddha et al. (2014). Indeed, our results revealed that the seed weight obtained after pollination by bees was 90 times higher than that obtained without bees. Similar results were also observed by Zdzislaw et al (2004) who found that the number of seeds produced by self-pollination was 10 times less than that produced by cross-pollination. According to Vaissière (2005), pollination by bees contributes 66% of seed production in a variety whose self-fertile plants produce pollen.

In conclusion, in the Tizi-Ouzou region, as in Cameroon, India and Pakistan, A. mellifera remains the main pollinator of A. cepa, pollinating 100% of visits, a number of visits which represents more than 45% of the total number of visits observed. L. malachurum and L. pauxillum are also very effective pollinators of this plant.

It is therefore suggested that nesting sites in areas close to crops should be conserved and protected, and that the unjustified use of pesticides during the flowering period should be avoided. These apoids can undoubtedly play a significant role in pollinating crops where their density is fairly high.

1. Lettuce (Lactuca sativa L.)

1.1. Flowering of the plant

Flowering began on 14 June 2016 and 08 June 2017 and lasted 26 and 24 days respectively. Lettuce is characterised by a capitulum inflorescence, with one plant bearing between 10 and 25 inflorescences. Colonisation of the plot by bees is noted as soon as the first lettuce flowers appear. The number of bees gradually increases over time, as the number of flowers increases. The highest rates of visits were recorded during the period of full flowering of this plant between 26 June 2016 and 28 June 2017 (Fig.25). Towards the end of flowering, the density of bees gradually decreases as flowering declines.

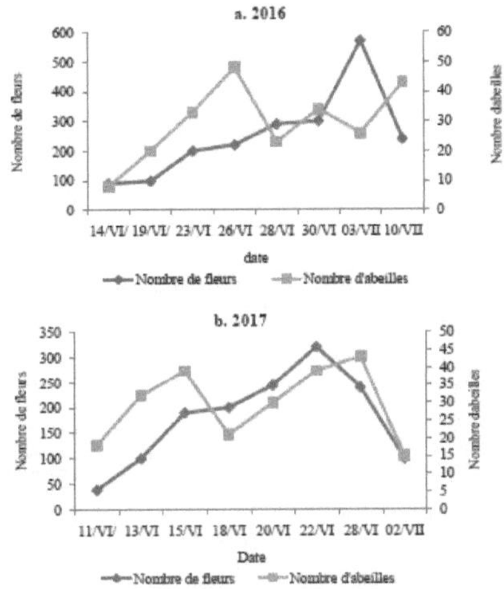

Figure 25: Flowering cycle of Lactuca sativa L. and bee density during the 2016 (a) and 2017 (b) flowerings.

4.2. Diversity and density of pollinators

The study carried out on Lactuca sativa during the two flowerings revealed that this plant attracts a pollinating entomofauna composed almost exclusively of apoid Hymenoptera belonging to two families: Halictidae and Apidae. Six species of bee have been recorded on the plant and are present during both flowerings (Tab.21). The Halictidae family is the most abundant on lettuce flowers, and two species of the Lasioglossum genus, L. malachurum and L. pauxillum, recorded the highest numbers of visits, accounting for more than half the total number of bees (74% and 78%) and a density of 5 to 8 individuals per 100 flowers (Tab.22). As for Apis mellifera and Eucera eucnemidea (Apidae), they recorded a low presence on the plant. Finally, the presence of L. smaragdellus and H. scabiosae is only episodic.

Table 21: Number of specimens (Nind) and relative frequencies (%Nind) of apoids recorded on transects during the 2016 and 2017 blooms.

Family	Species	N		N%	
		2016	2017	2016	2017
Halictidae	L. malachurum	95	96	42	40
	L. pauxillum	72	92	32	38
	L. smaragdellus	3	3	1	1
	Halictus scabiosae	7	5	3	2
Apidae	A. mellifera	18	16	8	7
	E. eucnemidea	33	29	14	12
Total	6 species	228	241	100	100

Table 22: Average number of pollinators per 100 flowers of Lactuca sativa during the 2016 and 2017 flowering seasons

Family	Species	2016	2017
	L. malachurum Kirby, 1802	6	8
Halictidae	L.pauxillum Schenck, 1853 Lasioglossum smaragdellus	5 <1	7 <1
	Halictus scabiosae Rossi, 1790	<1	<1
Apidae	A. mellifera L., 1758	1	2
	Eucera eucnemidea Dours, 1873	3	3

4.3. Foraging activity of apoids

During the two L. sativa flowerings in 2016 and 2017, the Apoids began their activity at 7am. Visits by these bees to the flowers are most frequent during the second and third observation periods (8am and 9am) when a maximum number of flowers are in bloom. From 10 o'clock onwards, the bees' visits diminish as many flowers begin to close (fig. 26).

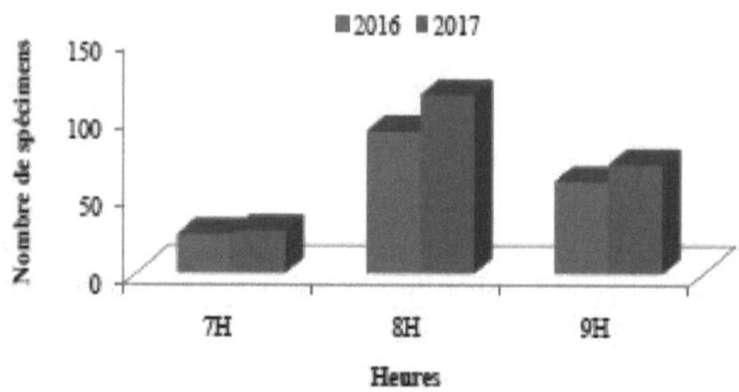

Figure 26: Total number of apoids recorded during the 2016 and 2017 blooms.

4.4. Foraging behaviour and efficiency pollinator

The foraging behaviour of the two most frequent species on the flowers, L. malachurum and L. pauxillum, is noted during flowering in 2016 and 2017. The majority of visits by these two Halictidae involve pollen collection, while the rest of the visits are devoted to nectar collection or the mixed collection of the two products (Tab. 23). These bees always make frontal visits to collect nectar, so these visits can all be fertilising.

Table 23. Distribution of visits (in %) for the two most abundant species on L. sativa according to the floral product harvested and pollination efficiency of these visits during flowering in 2016 and 2017. (P: pollen, N: nectar, +: pollinating visit, -: non-pollinating visit)

	L. pauxillum		L. malachurum	
	2016	2017	2016	2017
Number of visits	41	82	78	94
P+	83	60	67	49
N+	7	21	33	32
(P+N)+	10	19	0	19
Total pollinating visits	100%	100%	100%	100%

To study pollination efficiency, we assessed the foraging speed of L.malachurum and L.pauxillum. To do this, we measured the number of flowers visited in one minute by these bees. The results (Tab.24) show that during the 2016 flowering period, L.malachurum (8.9 ± 2.09 flowers/minute) was significantly faster than L.pauxillum (3.1 ± 0.92 flowers/minute) (Mann Whitney test $p<0.05$).

Table 24: Foraging speed of L.malachurum and L.pauxillum on L.sativa flowers during flowering in 2016.(Mann Whitney test (U); $p<0.05$; ±: standard error).

Species	number of specimens observed	Foraging speed	U	P
L.malachurumL.pauxillum	30 30	8.9 ± 2.09 3.1 ± 0.92	0	0

4.5. Searching for food bees

Observations were carried out during the 2017 flowering period to understand the type of foraging that bees encountered on L. sativa flowers were interested in. For the two most frequent bees on lettuce flowers, the majority of floral visits were for pollen collection (Fig.27 a and b). L. pauxillum recorded the highest proportions of pollen collection between 8 a.m. and 9 a.m., whereas nectar was mainly collected by this species during the first hour of observation. As for L. malachurum, it collected pollen in abundance at 7 am and 8 am, and nectar was collected at all observation times, but the highest proportions were observed at 9 am. In addition, mixed harvesting of the two products was observed in both species. L. pauxillum collects both floral products throughout its activity (from 7am to 9am). L. malachurum collects both floral products from 8 am onwards.

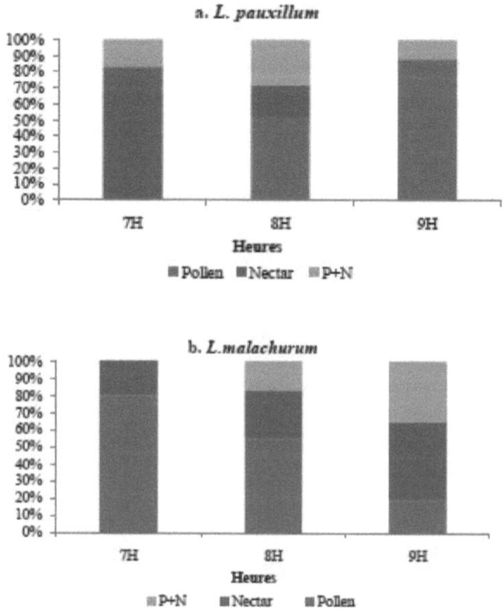

Figure 27: Percentages of floral products harvested by L. pauxillum (a) and L. malachurum (b).(b) on L. sativa at different times of day (2017 flowering)

4.6. Discussion

Studies carried out on L. sativa during the 2016 and 2017 flowering periods revealed that the main visitors to this plant are apoid Hymenoptera belonging to two families: Halictidae and Apidae. The most frequent species on the flowers are the two Halictidae, L. malachurum and L. pauxillum, which alone account for more than 70% of the visits observed. The Apidae are represented by A. mellifera and E. eucnemidea, which account for 14% and 12% of visits respectively. Our results concur with those of Goubara and Takasaki (2003) who found that in Japan, Lasioglossum villosulum trichopse is the most frequent species on L. sativa flowers.
Our study shows that L. smaragdellus and H. scabiosae are very poorly represented. This low density may be due to the attractiveness of other flowering plants in the vicinity of the lettuce crop.We also noticed that bees visited lettuce more frequently during the second and third observation periods (8am and 9am) when as many flowers as possible were in bloom. This activity can be explained by the fact that all the florets on a head open on the same day, early in the morning, and close shortly afterwards, never to open again. In some cases, they are open for only half an hour (Purseglove 1968, Jones and Rosa 1928, Thompson 1933 cited by Mc Gregor, 1976), but remain open longer in cool, cloudy weather. Our results are in line with those of Goubara and Takasaki (2003) in Japan, who found that Halictidae forage on lettuce flowers from 7am until midday.In addition, observation of the foraging behaviour of L. malachurum and L. pauxillum shows that these two species always effect a positive foraging whatever the product harvested, so all their visits can be fertilising. These bees when settling on the flower head, they hold the style bundle in their legs by leaning their body

outwards, which allows the release of large quantities of pollen seeds. This same behaviour was observed by Goubara and Takasaki (2004) on Lasioglossum villosulum trichopse foraging on lettuce flowers.The daily activity of bees on fleurs depends on the production of either pollen (Stone et al. 1998) or nectar (Pierre et al., 1996; Suzo et al., 2001; Pouvreau, 2004) during the day. Our observations in 2017 show that the majority of visits by L. malachurum and L. pauxillum involve pollen collection, with the remainder of visits devoted to nectar collection and mixed collections of the two products. Jones (1927) and Thompson (1933) also stated that Agapostemon texanus, Californicus crawford and Halictus spp. bees mainly collect pollen from L. sativa flowers.

2. Courgette (Cucurbita pepo L.)

2.1. Flower production and sex- ratio

Cucurbita pepo flowering started on 12 May 2016 and 18 May 2017 and lasted 70 and 38 days respectively. The total production of staminate (male) and pistillate (female) flowers during the two flowerings is shown in Figure 28 (a and b).

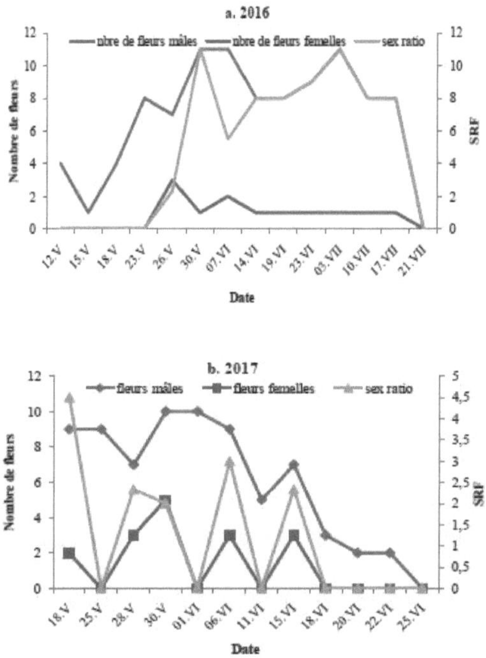

Figure 28: Production of male and female flowers by Cucurbita pepo plants during the growing season. 2016 (a) and 2017 (b) flowerings and floral sex ratio (SRF: ♂/♀).

The number of male flowers is always greater than the number of female flowers. In 2016, the start of flowering was characterised by a gradual increase in the number of staminate flowers, which stabilised in the middle of the cycle before decreasing at the end of the season. Female flowers appeared after the fourth day of observation and their number remained almost stable throughout the plant's flowering period. As for the 2017 flowering, male and female flowers appeared at the same time and their numbers varied constantly throughout the flowering period. The pistillate flowers disappeared before the staminate flowers. he sex ratio of the two populations also varies over time, with values of averages are: 7 ♂ / 1 ♀ in 2016 and 6 ♂ / 1 ♀ in 2017.

5.2. Diversity and density of pollinators

Observations made during the two flowering periods showed that the insects foraging on C. pepo flowers are mainly apoïdes Hymenoptera belonging to two families: Apidae and Halictidae.Six bee species were recorded on courgette (Tab.25). The species A. mellifera, Bombus terrestris (Apidae) and Lasioglossum villosulum (Halictidae) were present during both flowering periods. The honey bee recorded the highest number of visits. However, the bee densities per 100 flowers are overestimated because the foragers are scattered over only a few flowers. The Apidae Eucera eucnemidea and the two Halictidae: L. malachurum and L. glabriusculum were only observed during flowering in 2016.

Table 25. Average densities per 100 flowers of Cucurbita pepo pollinators during the 2016 and 2017 flowering seasons. (-: absence).

Family	Species	2016	2017
Apidae	A.mellifera (L. 1758)	37	73
	B. terrestris (L. 1758)	1	1
	E.eucnemidea (Dours, 1873)	1	-
Halictidae	L. malachurum(Kirby 1802)	5	-
	L. villosulum (kirby 1802)	5	3
	L. glabriusculum (Morawitz, 1872)	1	-

5.3. Trend in the number of visits by honey bees during the two flowering periods

In 2016, the peak in honeybee visits was recorded on 30 May (Fig. 29a) and the trend in the number of honeybee visits roughly follows the shape of the floral curve. During the 2017 flowering period, the pattern of visits also roughly follows that of flower development (Fig. 29b). Bee visits are more numerous at the start of flowering, peaking in the middle of the flowering period.

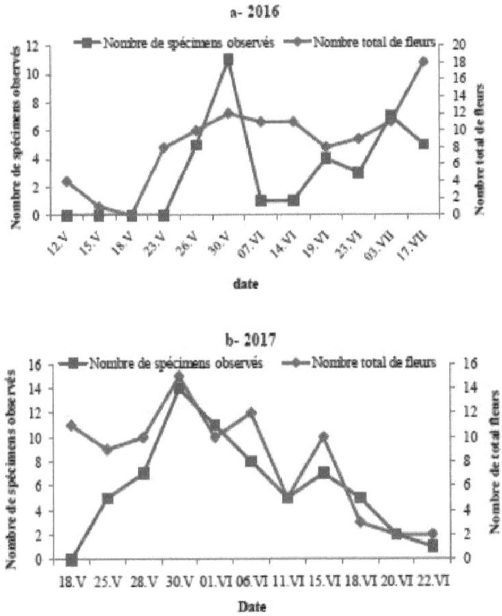

Figure 29: Trend in the number of visits by Apis mellifera to Cucurbita pepo in relation to flower density during the 2016 (a) and 2017 (b) flowering seasons.

5.4. Foraging behaviour and efficiency pollinator

5.4.1. Foraging activity

Courgette flowers have a short lifespan, with each female flower having around 2 to 3 hours in which to be pollinated. Visits by honey bees to flowers are most frequent during the second and third observation periods (8am and 9am) when a maximum number of flowers are in bloom. From 10 a.m. onwards, visits by the bees decrease as many flowers begin to close (Fig. 30).

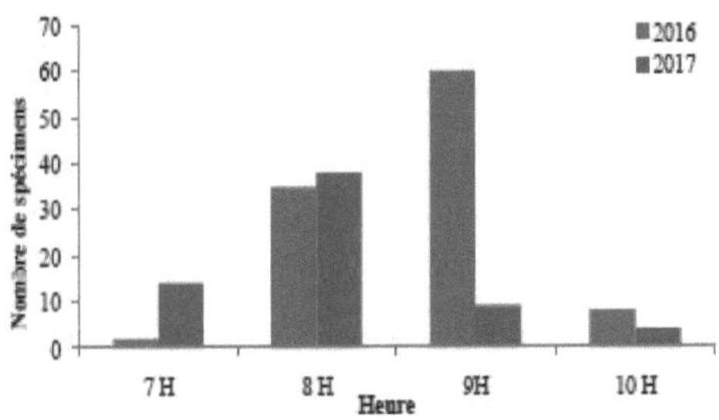

Figure 30: Number of Apis mellifera visits recorded on C. pepo at different observation times during the 2016 and 2017 flowering seasons

5.4.2. Foraging behaviour

Observations of foraging activity on courgette by honey bees during the two successive years of the study revealed high rates of nectar-harvesting visits by honey bees, with rates of 81% and 86% respectively for the 2016 and 2017 flowering seasons. The rest of the visits (19% in 2016 and 14% in 2017) were devoted to pollen collection. Most visits (93%) by wild bees are also devoted to collecting nectar. Although female flowers produce more nectar than male flowers, bees prefer staminate flowers (Tab. 26). On female flowers, before penetrating to the bottom of the corolla to collect nectar, the honey bee first lands on the stigma, where it moves around. Sometimes they do this before leaving the flower. In this way, the honey bee's visits to pistillate flowers are all potentially pollinating. On male flowers, the bee lands on the anthers, which it scratches with its front legs to collect the pollen. To collect the nectar, the insect lands near the edge of the corolla and heads for the bottom. On leaving the flower, the bee emerges covered in pollen. She spends a lot of time brushing herself and the pollen often falls in small piles.

Table 26: Distribution (in %) of visits by A. mellifera and wild bees to C. pepo according to the floral product harvested during the 2016 and 2017 flowering seasons. +: pollinating visit; n= number of specimens observed; *= n (nectar+ pollen) of male flowers

Flowering	2016								2017			
Species	Apis mellifera				Wild bees				Apis mellifera			
harvested product	n		%		n		%		n		%	
Nectar + (female)	22		21		18		32		22		28	
Nectar (male)	63	83*	60	79*	34	38*	61	68*	46	55*	60	72*
Pollen (male)	20		19		4		7		9		12	
Total number of visits	105		100		56		100		77		100	

5.4.3. Efficiency pollinator

The pollination efficiency of bees on flowers depends on the time the insect spends on the two types of flowers. If a species spends a long time visiting flowers, the rate of its pollinating visits would be small. We therefore measured the time spent by the honey bee on the two types of flowers.The results (Tab.27) show that during flowering in 2016 and 2017, the bee spent significantly more time on pistillate flowers than on staminate flowers. (Mann Whitney test (U); p<0, 05).

Table 27: Duration of visits by A. mellifera to C. pepo flowers during flowering in 2016 and 2017. ±: standard error; n = number of specimens observed. (Mann Whitney test (U); p<0, 05).

Flowering	Visiting time on male flowers (in seconds)	Duration of visit to female flowers (in seconds)	U	P
2016	26.5 ±10.67 (n=30)	46.66 ± 15.61 (n=30)	140	0
2017	26 ± 10.61 (n=30)	50 ± 14.38 (n=30)	95	0

5.5. Foraging for honey

The honey bees' floral visits during the two flowering periods were devoted exclusively to collecting nectar (Fig.31). Pollen was also collected at 8 and 9 o'clock, but only in small quantities.

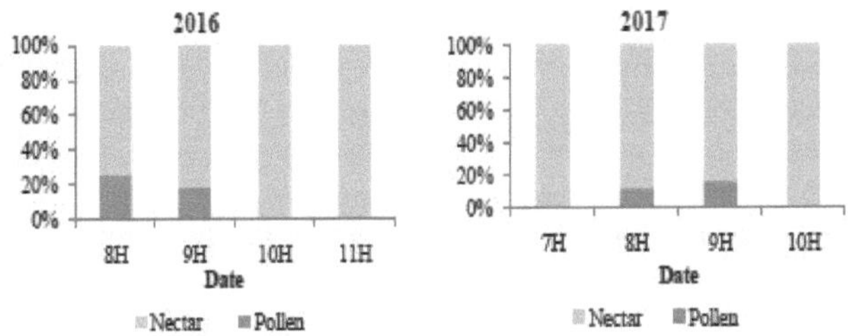

Figure 31: Apis mellifera foraging on Cucurbita pepo at different times of day during the 2016 and 2017 flowering seasons.

5.6. Discussion

Studies carried out on C. pepo during the two flowering periods showed that the honey bee is the most abundant species on this plant. Studies carried out in Algeria (Benachour, 2008), Italy (Nepi and Pacini, 1993), Brazil (Krug et al., 2010) and the United States (Esther Julier and Roulston, 2009) have also shown that A. mellifera is the most abundant visitor to courgette. In India, Devika Rani et al (2017) revealed the presence of four species of the Apis genus on C. pepo. In the United States, Cane et al. (2010) recorded the solitary bee Peponapis pruinosa from the family Apidae as the most abundant species on this plant. Visits by honey bees to flowers are most frequent between 8am and 9am, when a maximum number of flowers are in bloom. From 10 a.m. onwards, visits by the bees decrease as many flowers begin to close. Our results are similar to those of Sanduleac (1959), Benachour (2008) and Mudssar et al. (2014) who reported intense honeybee visits between 8am and 9am. This can be explained by the fact that the flowers Squash trees generally open shortly after dawn and close permanently at midday or shortly afterwards, often receiving frequent early morning visits from pollinators (Tepedino, 1981). In fact, this period corresponds to maximum pollen and nectar production, resulting in intense bee activity. This high level of bee activity ensures adequate pollination of the plant, since the flowers, like those of other Cucurbitaceae, remain open for only one day and for only a few hours.Our observations showed that the majority of visits by A. mellifera are devoted to nectar collection. Our results are in line with those of Benachour (2008) in Constantine, where she reported that visits to flowers by honey bees are exclusively for nectar collection. Nectar is secreted in large quantities by flowers (Tepedino 1981; Nepi & Pacini 1993). Pollen, on the other hand, is rarely collected by the bee. Staminate flowers do not produce much pollen and it is most abundant during the first hours of anthesis (Nepi & Pacini 1993). Michelbacher et al (1964) and Philippe (1991) also noted that this product is rarely collected by bees. However, Devika Rani et al (2017) revealed that more than 47% of honey bee visits are devoted to pollen harvesting and 19% to nectar harvesting.The pollination efficiency of bees is measured by the distribution of individuals on male and female flowers, and the number of flowers foraged per minute. Our observations showed that honey bees visit male flowers more often than female flowers; this could be due to the fact that their number is always higher than that of female flowers. Our results corroborate those of Benachour (2008) and Mudssar et al. (2014) who explained this by the fact that the sugar concentration of the nectar produced by male flowers is higher. However, in courgette, it has been shown that female flowers produce more nectar than male flowers and are more frequently visited by bees (Tepedino 1981; Couto et al. 1990; Philippe 1991; Nepi & Pacini 1993).Visits by honey bees are also faster on staminate flowers. These results are similar to those obtained by Tepedino (1981), Nepi & pacini (1993) and Benachour (2008). According to the latter authors, bee visits are slower on pistilated flowers due to the higher quantity of nectar and the position of the nectary.

3. Watermelon (Citrullus lanatus)

3.1. Flower production and sex ratio :

Citrullus lanatus flowering started on 04 July 2016 and 02 July 2017 and spanned 40 and 45 days respectively. The total production of staminate (male) and pistillate (female) flowers during the two flowerings is shown in Figure 32 (a and b).

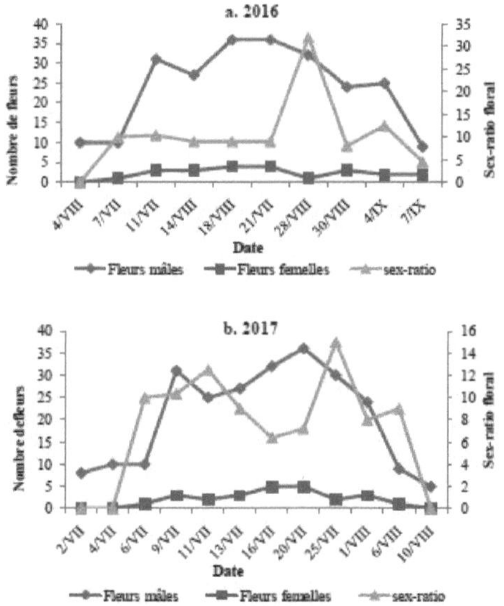

Figure 32: Production of male and female flowers by Citrullus lanatus plants during the growing season. 2016 (a) and 2017 (b) flowerings and floral sex ratio (♂/♀)

The male flowers appear a few days before the female flowers and their number is always higher than that of the female flowers. This number varies constantly, being low at the beginning and end of flowering, and peaking on 18 and 21 July 2016 and 20 July 2017. Similarly, the number of female flowers fluctuates during flowering. The sex ratio of the two populations also varies over time. In 2016, it was stable at the start of flowering and varied constantly at the end of the season. The average sex ratios of the population are: 24 ♂ / 2 ♀ and 21 ♂/ 2 ♀ respectively in 2016 and 2017.

6.2. Diversity and density of pollinators

Observations carried out on the flowers of C. lanatus during the two floraison periods show that the foraging insects of this plant are all apoïdes Hymenoptera belonging to three families. Eleven species were recorded, including 6 Apidae, 4 Halictidae and 1 Megachilidae (Tab.28). Among the species encountered, Laioglossum pauxillum belonging to the family Halictidae, is more frequent and its density per 100 flowers is higher during both flowerings. Among the Apidae, A. mellifera and Nomada sp. are the most abundant on the flowers, with respective densities per 100 flowers of 30 and 35 individuals in 2016 and 22 and 34 individuals in 2017. Ceratina saundersi (Apidae) is also relatively well represented. The Megachilidae family was represented by a single species during both flowering periods. The other species made occasional visits. (Tab. 29).

Table 28: Number of specimens (Nind) and relative frequencies (%Nind) of apoids recorded on transects during the 2016 and 2017 blooms.

Family	Species	2016		2017	
		N	N%	N	N%
	Apis mellifera	66	29.07	71	26.59
	Nomada sp	49	21.59	63	23.60
Apidae	Ceratina saundersi Ceratina cucurbitina	22 1	9.69 0.44	19 17	7.12 6.37
	Amegella sp.1	1	0.44	5	1.87
	Amegella sp.2	2	0.88	-	-
	Laioglossum pauxillum	69	30.40	75	28.09
Halictidae	Lasioglossum sp. Halictus scabiosae	1 5	0.44 2.20	- 10	- 3.75
	Halictus sp.	3	1.32	-	-
Megachilidae	Megachile sp.	8	3.52	7	2.62
Total	11 species	277	100	267	100

Table 29. Average densities per 100 flowers of C. lanatus pollinators during the 2016 and 2017 flowering seasons. (- : absence)

Family	Species	2016	2017
	A. mellifera	30	35
	Nomada sp.	22	34
Apidae	Ceratina saundersi Ceratina cucurbitina	9 4	11 10
	Amegella sp1	3	4
	Amegella sp2	3	-
	L. pauxillum	34	29
Halictidae	L. glabriusculum H. scabiosae	1 4	- 6
	Halictus sp.	2	-
Megachilidae	Megachile sp.	8	6

6.3. Number of visits by bees during the two flowering periods

In 2016, the peak of bee visits was recorded in the middle of the flowering period (18 and 21 August) (fig.33 a) and the evolution of the number of bee visits roughly follows the shape of the flowering curve. During the 2017 flowering season, the peak in visits was recorded towards the end of the flowering period (20 July) (fig. 33 b). From 20 July onwards, the pattern of visits also roughly follows the pattern of flowering.

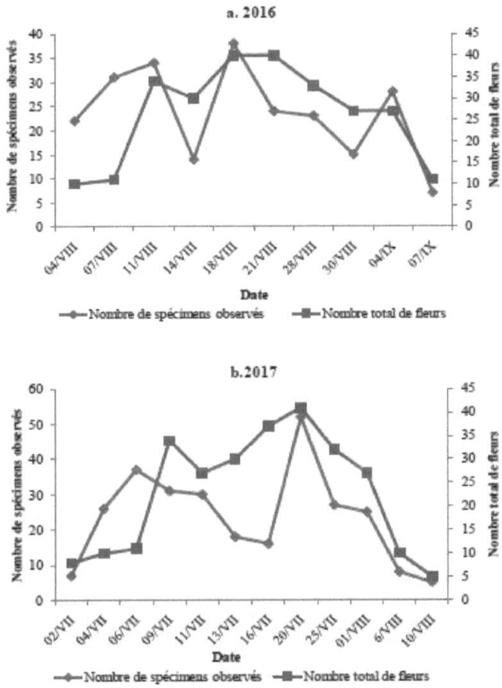

Figure 33: Trend in the number of visits by bees to C. lanatus in relation to the number of flowers during the 2016 (a) and 2017 (b) flowering seasons.

6.4. Foraging activity and behaviour of apoids

During the two watermelon flowering periods, the daily activity of the the most abundant pollinators on the flowers.The different species show varying patterns of daily distribution. During the 2016 flowering period (Fig. 34a), the four bee species monitored began to visit the watermelon flowers from 8am onwards, with the honey bee making the most visits at this time of day, its activity ceasing at 10am. The activity of the other species intensifies at 9am and then drops off or disappears at 10am. In 2017, bee visits began timidly at 7am and intensified at 8am and 9am, declining at 10am (Fig. 34b). A study of the behaviour of bees observed on the plant did not reveal any pollen collection by these insects. Nectar was the only product collected by all the species that visited the plant.

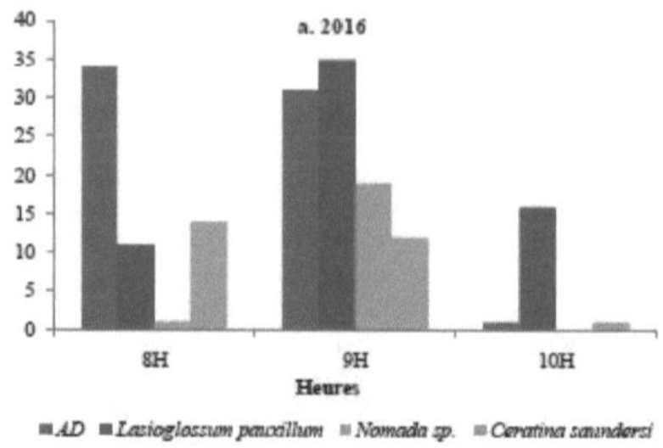

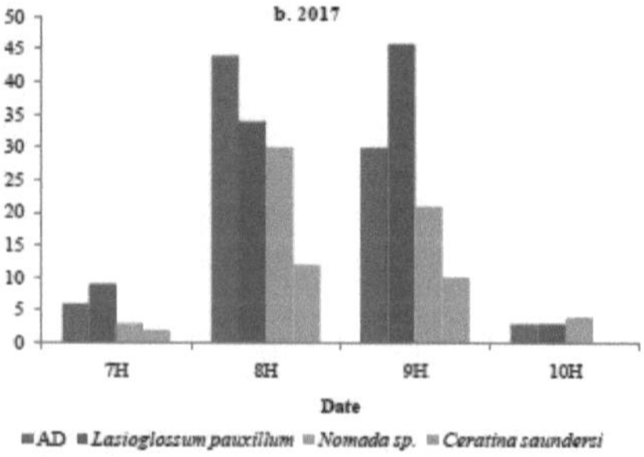

Figure 34: Changes in the number of main Citrullus lanatus pollinators at different times of day during flowering in 2016 (a) and 2017 (b).

6.5. Pollination efficiency of bees

The foraging speed of the main species found on C. lanatus flowers was recorded during flowering in 2016 (Tab. 30). Visits by honey bees were the most frequent, while L. pauxillum recorded the lowest speed on watermelon flowers.

Table 30: Foraging speed of bee species on Citrullus lanatus during the 2016 flowering season.(N = number of specimens observed; ±: standard error)

	N	Foraging speed (flowers/min)
Apis mellifera	30	9.9 ± 1.25
Ceratina saundersi	30	8.4 ± 2.7
Nomada sp.	30	6.3 ± 1.9
Asioglossum pauxillum	30	4.9 ± 0.83

Like courgette, watermelon is a monoecious plant and the pollination efficiency of the species can also be measured by the proportion of visits to staminate flowers and pistillate flowers, and by the time spent by bees on each type of flower. The bees showed a clear preference for staminate flowers (Tab.31) and their visits to pistillate flowers are all potentially fertilising since they come into contact with the stigma.

Table 31. Percentages of visits made by honey bees and wild bees to male (♂) and female (♀) flowers of C.lanatus (2016 and 2017 flowering seasons). (N = total number of specimens observed; n = number of specimens observed on each flower type).

Flowering	2016					2017				
Flower		♂		♀			♂		♀	
Species	N	n	%	n	%	N	n	%	n	%
Apis mellifera	66	52	79	14	21	71	54	76	17	22
Nomada sp.	49	34	69	15	31	63	58	92	5	5
Lasioglossum pauxillum	69	50	72	19	28	75	65	87	10	11
Ceratina saundersi	22	14	64	8	36	19	15	79	4	5

The time spent by bees on male and female flowers was measured during flowering in 2016 (Tab.32). We found that the four bee species observed spent significantly more time on female flowers than on male flowers (Mann Whitney test (U); p<0, 05).

Table 32: Average time (in seconds) spent by the different species of bees on the male (♂) and female (♀) flowers of C. lanatus during flowering in 2016. (n = number of specimens observed ; ± : standard error (Mann Whitney test (U); p<0, 05)

	Apis mellifera	Ceratina saundersi	Nomada sp.	Lasioglossum pauxillum
Flowers ♂	(n=19) 6.03±0.89	(n=15) 8.9±2.58	(n=22) 10.12±3.06	(n=22) 8.84±1.33
Flowers ♀	(n=19) 11.66±2.79	(n=15) 18.66±5.16	(n=22) 12.72±2.14	(n=22) 17.31±3.15
U	0	0	135	0
P	0	0	0.01	0

6.6. Discussion

The study carried out on C. lanatus during the two flowering periods showed that the flowers of this Cucurbitaceae attract numerous visitors, the majority of which are apoid Hymenoptera. 3 families, 7 genera and 11 species of bees were recorded on the plant. Apidae, represented mainly by A. mellifera and Nomada sp., are the most abundant foragers (on average 50% of visits observed) on the flowers during both flowering periods.

Among the Halictidae, L. pauxillum recorded the highest number of visits. This species accounted for an average of 88% of visits by Halictidae. As for the Megachilidae, they are the least abundant visitors to the plant (3% of visits observed).

Several authors (Rao and Suryanarayana, 1988; Sanford, 1992; Hodges and Baxendale, 1995; Stanghellini et al., 1998; Njoroge et al., 2004; Taha and Bayoumi, 2009 and Azo'o et al., 2010) have shown that the honeybee is the main pollinator of watermelon. However, our study identified other important bees on this plant. These results are consistent with those of Njoroge et al (2010), who recorded visits by 3 species of Lasioglossum bees to watermelon flowers.Bees are most active in the morning, with peak numbers between 8 and 9 am. This intense morning activity is synchronised with the maximum opening of the flowers, which begins one to two hours after sunrise (Mc Gregor, 1976). According to Poole and Porter (1933) and Sedgley and Buttrose (1978), the stigma is receptive throughout the day, but pollination takes place mainly in the morning. These observations are similar to those of Njoroge et al (2004) in Kenya and Taha and Bayoumi (2009) in Egypt.

Collecting nectar is the objective of all the bees that visited the plant during our observations. This corroborates the results obtained by Njoroge et al (2009), Taha and Bayoumi (2009) and Azo'o et al (2010) who also found that nectar is the main product harvested by bees visiting watermelon.

Although visits by bees to pistillate flowers are potentially pollinating, they are much less numerous than those to male flowers. Collison and Martin (1979) also found that honey bees visited staminate flowers more frequently. The difference in nectar secretion between the two types of flower may explain the bees' preferences for one or the other. However, it has been shown that pistillate flowers produce more nectar, which is more concentrated in sugars (Taha and Bayoumi, 2009). By The greater attraction of staminate flowers to bees can therefore be explained by the higher number of male flowers compared with female flowers.

The pollination efficiency of bees is measured by the distribution of individuals on male and female flowers, and the number of flowers foraged per minute. Our observations showed that the honey bee is the fastest species, visiting watermelon flowers at a rate of 9.9 ± 1.25 flr/min. These results are similar to those found by Azo'o et al (2010) who noted a speed of 10.20 ± 2.75 flr/min for the honey bee. As for the time spent by bees on male and female flowers, it seems that the less time a bee spends on pistillate flowers, the greater the rate of its pollinating visits (Tepedino 1981; Cervancia & Bergonia 1991). Our results showed that the bees spent more time on female flowers than on male flowers. The abundance of nectar in female flowers may explain the slowness of foragers' visits to these flowers. Indeed, Collison & Martin (1979) found that the duration of visits to flowers was correlated with the quantity of nectar present. The authors noted that if a honey bee is not disturbed, it will remain on the flower until it has collected all the nectar.

7. Orange (Citrus sinensis)

7.1. Flowering

The floraison started on 02.March.2016, 05.April.2017 and 27 March 2018 and lasted 58, 23 and 44 days, respectively (Fig. 35a, b and c). The floraison was more reduced in 2017 due to a drier winter and a very hot summer with scorching temperatures compared with 2016 and 2018. Maximum flowering was recorded on 21 April 2016, 16 and 19 April 2017 and 09 April 2018. The orange blossoms were open throughout the day.

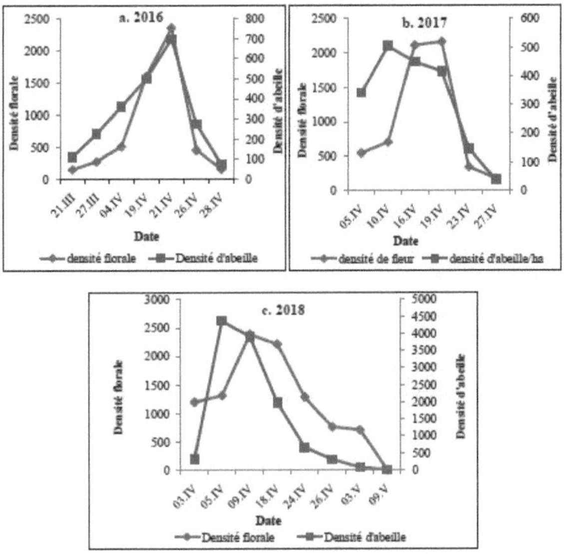

Figure 35: Citrus sinensis flowering cycle and bee density in 2016 (a), 2017 (b) and 2018 (c).

The colonisation of the orange orchard by bees is noted as soon as the first flowers appear. The density of bees follows the evolution of the flowers. The highest visit rates were recorded during the orange's peak flowering period on 21 April 2016, 10 April 2017 and 5 April 2018. Towards the end of the flowering period, the number of bees gradually decreases as the flowering declines.

7.2. Diversity and density of pollinators

The study carried out on C. sinensis showed that the flowers of this citrus fruit attract a pollinating entomofauna composed almost exclusively of apoid Hymenoptera, all belonging to the Apidae family. Three species have been identified: A. mellifera, X. violacea L. 1758 and B. terrestris L. 1758. The honey bee is observed throughout the year. It is by far the most dominant species, with densities exceeding 300 individuals/ha, and is relatively very frequent, particularly during the 2018 observations. As for the other two species, their presence is only episodic (Tab. 33).

Table 33: Pollinator density per hectare in the orange orchard.

Year	Species	Density/ha
	A. mellifera	321
2016	X. violacea	5
	B.s terrestris	2
	A. mellifera	316
2017	X. violacea	1
	B.s terrestris	2
	A. mellifera	1755
2018	X. violacea	2
	B.s terrestris	<1

7.3. Foraging activity and behaviour of apoids

The daily activity and foraging behaviour of A. mellifera were observed during the three orange blossom periods. The honey bee began foraging on the orange flowers from the first hour of observation (8 a.m.), its activity intensified at 10 a.m. and 11 a.m. and then gradually decreased in the afternoon (Fig. 36). The study of honeybee behaviour on orange flowers did not reveal any pollen collection. All visits are devoted to nectar collection, with the bee collecting nectar from the basal part of the style and the stamens. Some bees collect nectar even before the flowers open while trying to open them.

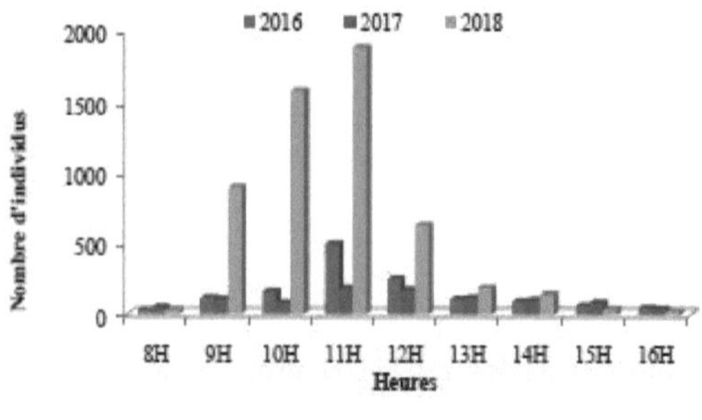

Figure 36: Changes in the number of honey bees on flowers of Citrus sinensis during observations in 2016, 2017 and 2018.

7.4. Efficiency pollinator

Foraging speed seems to be the main parameter for estimating pollination efficiency. In order to determine whether honeybees forage on orange flowers at the same rate throughout the day, we measured their foraging speed at 8 am, 10 am and 4 pm (Tab.34). In addition, a variance test was carried out to determine the effect of the time of activity on the foraging speed of the honey bee. The analysis of variance (Tab.35) at the 5% threshold shows that the calculated probability (P= 0) is less than $\alpha = 0.05$, so there is a significant difference in the foraging speed of the honey bee between the three hours of activity. It is faster at 8 am and 4 pm and slower at 10 am. On the other hand, we compared the foraging speed of honey bees on three citrus species (Tab.36): mandarin (C. reticulata), orange (C. sinensis) and lemon (C. limon). The results showed that the bee is clearly faster on mandarin flowers (7 flrs/min). As for the lemon and orange, they forage at roughly the same speed (3 flrs/min and 2.8 flrs/min respectively).

Table 34: Foraging speed of Apis mellifera on Citrus sinensis flowers during the 2016 floraison.

Hours	Number of individuals observed	Speed (flowers/min)
8 H	30	4.5 ± 1.04
10 H	30	2.7 ± 1.5
16 H	30	3.9 ± 0.75

Table 35: Results of the analysis of variance for Apis mellifera observed in the Tizi-Ouzou region

	S.C.E	DDL	C.M.	TEST F	PROBA
TOTAL VAR.	158.48 9	89	1.781		
VAR.time	44.956	2	22.478	17.225	0
VAR.RESIDUAL 1	113.53 3	87	1.305		

Table 36: Foraging speed of Apis mellifera on Citrus flowers during the 2016 floraison.

tree	Number of individuals observed	Speed (flowers/min)
Citrus sinensis	30	2.8 ± 1.5
Citrus reticulata	30	7 ± 1.9
Citrus limon	30	3 ± 0.9

7.5. Discussion

The study carried out on C. sinensis during the 2016, 2017 and 2018 flowerings revealed that this fruit tree attracts many visitors, the majority of which are apoid Hymenoptera of the Apidae family. The honey bee A. mellifera was the most abundant species on the fleurs over the three study periods. These results are consistent with those of several authors who have studied citrus pollination. In Algeria, Barbier (1964) only recorded the presence of the honey bee on Clementine flowers and noted that this insect is undoubtedly the main citrus pollinator. The same applies to studies carried out in Brazil (Malerbo-Souza et al., 2004 and Ribeiro et al., 2017), Mexico (Grajales-Conesa et al., 2013) and India (Singh, 2016), which revealed that A. mellifera is the most abundant species on orange flowers. As for B. terrestris and X. violacea, their presence on the flowers is only episodic. This low density may be due to the attractiveness of other flowering plants in the vicinity of the orchard. According to Barbier (1964), the lack of pollinators other than bees may be due to the clayey nature of the soil or its humidity.Visits by honeybees to orange blossoms were noted as soon as the first flowers opened. These visits are very high at 11am, and this intense activity may be explained by a high nectar secretion. In this respect, Ribeiro et al (2017) revealed a significant difference in the volume of nectar secreted by orange flowers during the hours of the day, revealing intense nectar secretion at 10:30 am. According to Vansell et al (1942), orange nectar is attractive when it is concentrated in sugar. This concentration is affected by the humidity of the air; in fact, nectar is rich in sugar in dry weather. The study of the foraging behaviour of A. mellifera did not reveal pollen collection and all its visits were for nectar collection. In general, citrus fruits are not considered an excellent source of pollen by beekeepers (Mc Gregor, 1976). Our results agree with those of Ribeiro et al. (2017) who reported that 90% of honey bee visits are devoted to nectar harvesting. As for Singh (2016), he revealed the presence of three bee species of the genus Apis on orange flowers collecting both nectar and pollen products. Some bees collect nectar even before the flowers open while trying to open it, this behaviour may be due to the secretion of nectar by the flower buds. The orange is one of the rare flowers that secrete nectar before the petals develop (Vansell et al., 1942).

With regard to the pollination efficiency of honeybees on orange flowers during the different observation times, our results showed that honeybees are more efficient at 08:00 and 16:00 and slower at 10:00. This difference in speed is perhaps explained by the fact that nectar is richer in sugar in dry weather, but also by the quantity of nectar secreted. If the bee finds that the flower it visits is rich in nectar, it will spend more time exploiting it. The results of work by Ribeiro et al (2017) on The orange studies point in the same direction, demonstrating that a large quantity of nectar is indeed secreted at 10.30am. In addition, the bees appear to be more active at other times of the day (8am and 4pm) when temperatures are supposed to be milder than at 10am.

As honeybees are the main pollinators of citrus fruit, wind and other insects play a minimal role. For this reason, the location of bee colonies has often been recommended. Vansell and Watkins (1942) reported that some flowers contained 1.5 bee loads of nectar, averaging 20 microlitres, compared with 0.8 to 2.4 microlitres per flower for alfalfa, another important source of nectar. Because of the large quantity and superior quality of honey produced by bees that feed on citrus blossom, many beekeepers set up their colonies in or near most orchards. On the other hand, to ensure good pollination of the flowers, citrus growers need to provide an appropriate number and location of bee colonies.

CONCLUSION AND OUTLOOK

Investigations carried out during 2016, 2017 and 2018 in the Tizi- Ouzou region on 6 market garden plants and the orange tree revealed that the honey bee is the main pollinator of courgette and orange tree. It is also the main visitor to watermelon, alongside four other solitary bee species: Ceratina saundersi, Ceratina cucurbitina, Nomada sp. and Lasioglossum pauxillum. In addition, two Halictidae, L. pauxillum and Lasioglossum malachurum, appear to be the main pollinators of lettuce and onions. The two legumes, fava beans and chickpeas, are mainly pollinated by other bee species. Bean flowers are highly prized by two species of the Apidae family, Eucera numida and Apis mellifera. Bombus terrestris and Xylocopa violacea (Apidae) also forage on the bean. For chickpeas, Eucera decolorata and Eucera eucnemidea (Apidae) are the main pollinators. Lithurgus chrysurus (Megachilidae) and Apis mellifera are also fairly frequent visitors to this legume.A study of the effect of entomophilic pollination on bean and onion production revealed that the presence of pollinating insects significantly improves the yield of these plants.With regard to the distribution of bees by family on all the plants studied, the Apidae, with an average of 65% of visits observed, are the most abundant apoids. The Halicidae come second with an average of 32% of visits observed.

The honey bee is the only species common to all these plants. It is the most efficient and widely used insect for pollination. Its foraging activity meets the nutritional needs of the colony, with optimised foraging (Alleaume, 2012). It has been domesticated for a very long time, and was already being bred by man in ancient Egypt (4,000 years ago), mainly for the production of honey and wax, rather than for any superior pollination skills (Batra, 1995 cited by Benachour, 2008).Until recently, the honey bee was considered to be one of the most important pollinators. It was credited with no less than 80% of pollination activity in agricultural crops. However, its importance may have been greatly exaggerated. In 2007, for example, the honeybee population in Great Britain was responsible for no more than a third of total pollination activity, with the remainder coming from wild pollinators, particularly wild bees. Another study clearly indicates that wild bees foraging on flowers increase fruiting in agricultural crops even when honey bees are abundant. As a recent large-scale study has shownin which the pollination activity of honey bees and wild foragers in 41 crops on all continents was compared. Indeed, on certain crops and under certain conditions, the honey bee is not efficient (Batra 1995 cited by Benachour, 2008). According to Tasei (1978 cited by Benachour, 2008), the bee collects nectar without triggering the flower, or only participates to a very small extent in pollination (nectar foragers are triggered at a rate of only 1%). Wild bees are generally the most efficient pollinators: a few hundred female Osmia cornuta mason bees are enough to pollinate a hectare of apple or almond trees. Certain flowers whose nectar is difficult to exploit and avoided by domestic bees, such as red clover, alfalfa or tomatoes, are visited by specialist wild bees (Pfiffner and Mûller, 2014). In Japan, Osmia cornifrons is bred to pollinate apple trees (Sekita &Yamada 1993 cited by Benachour, 2008). The pollination efficiency of these species is due to their foraging behaviour (loyalty to the host plant, rapid foraging, small foraging area, efficient pollen collection and distribution). Bumblebees, which are better pollinators than honeybees, are used in greenhouses to pollinate tomatoes. For better yields, this crop requires vibratory pollination, which cannot be achieved by honeybees.A study of the foraging behaviour of the wild bee Eucera numida on fava beans also shows the pollinating superiority of this species over the domestic bee. In addition, several species of wild bee fly even when temperatures and solar radiation are low and play an important role in pollination, particularly for fruit trees. However, compared with certain species of bumblebee, the honey bee does not cope as well with adverse weather conditions. They also adapt less well to greenhouse conditions than bumblebees and tend to turn away from the crop they are pollinating in favour of a more lucrative food source (Pouvreau, 2004).

A study of the pollination efficiency of certain bee species based on their foraging speed on the flowers of the plants studied revealed that the honey bee is the fastest on onion and watermelon flowers. On the other hand, male E. numida outperformed the honey bee on bean flowers. On chickpea flowers, E. decolorata is faster than E. eucnemidea. Similarly, on lettuce flowers, L. malachurum is by far faster than L. pauxillum. The main guarantee of effective pollination of wild and cultivated plants therefore lies in the combination of healthy populations of honey bees and large, species-rich communities of wild bees. According to Pfiffner and Mûller (2014), honeybees can simply complement the pollination activity of wild foragers, not replace it. In fact, contrary to expectations, it is not the number of foraging bees that proves to be the most important factor in the success of coffee pollination, but the diversity of bee species, so honey bees cannot be the only pollinators in ecosystems. Unfortunately, wild bee populations fluctuate to a greater or lesser extent from year to year. Observations made in certain regions indicate that while some species remain stable, others are becoming rarer, declining or even disappearing (Pouvreau, 2004).The main causes of decline are twofold: exposure to chemicals and loss of food resources. The first is exposure of bees to plant protection products or pesticides in cultivated areas. Bees are exposed directly during the application of treatments, but also via the residues of these chemicals contained in the matrices harvested by the bees (Anonymous 6, 2013). Pesticides, particularly neurotoxins, disorientate bees, modify their behaviour and weaken their immune systems. The second reason for the decline in bee numbers is the reduction in food resources. To ensure their life cycle, bees need quality pollen from a diverse flora (a source of protein) and nectar (a source of energy). The reduction in biodiversity in agricultural areas, linked in particular to monoculture, has resulted in a reduction in the number of plant species available and a shortening of their flowering time. In addition to the lack of pollen, which leads to a lack of sufficient reserves, there is a lack of diversity, which affects the health of bee populations (Anonymous 6, 2013).The modification of landscapes (Holzschuh et al, 2007, in Lachaud and Mahé, 2008) resulting from changes in farming, pastoral and forestry practices, the extreme decline in traditional forage crops (alfalfa, sainfoin, clover), and the regular use of selective herbicides in field crops, The systematic amendment of permanent grassland, which favours Poaceae to the detriment of entomophilous flowering plants (dicotyledons), completes the list of reasons for the decline in wild pollinator populations (Lachaud and Mahé, 2008). Rasmont et al, (2003) in addition to the weeding of field crops and the virtual disappearance of Fabaceae, cite the excessive "cleaning" of wasteland and roadsides. According to Lachaud and Mahé (2008), pollution masks some of the scent of flowers and the sexual pheromones of bees, making it more difficult for them to find food and partners for reproduction. The intensification of urbanisation, the creation of industrial zones and forest fires are also the causes of the disappearance of many biotopes that provide habitat and food for pollinating insects.The decline in bee populations (both domestic and wild) would lead to an impoverishment of both plant and animal biodiversity. Our agriculture and therefore our food security depend on this biodiversity. The decline in these populations of pollinating insects can have more or less serious effects on the survival of native plant species that require entomophilic pollination, with unfortunate economic consequences. Targeted protection measures can be effective in promoting wild bees. Generally speaking, the highest priority should be given to preserving habitats where flowers and small structures are abundant. Every measure aimed at increasing the quantity, diversity and improving the distribution of flowering plants and small structures with good exposure to the sun is beneficial for the diversity of species and the size of wild bee populations. In this context, close proximity between food resources and nesting sites and the permanent presence of flowers from early spring to late summer are of crucial importance (Pfiffner and Mûller, 2014). It is also advisable to maintain uncultivated plots within or close to cultivated land. These uncultivated strips not only provide a habitat for auxiliary insects (pollinators or predators of harmful insects) but can also increase the number of these insects on crops, thereby increasing yields (Francis et al. 2003; Morandin & Winston, 2006 cited by Benachour, 2008). Another way of protecting and increasing the population of wild pollinators is to select and breed the species best suited to effectively pollinating the flowers of cultivated plants. The breeding of wild bee species endemic to Algeria could be carried out as in Europe and North America to meet the pollination needs of plants that are not pollinated by the domestic bee. In conclusion, the observations made in the Tizi-Ouzou region on a number of vegetable plants show that despite the major role played by the honeybee in the wild bees remain interesting pollinating agents. It can therefore be said that these bees play a complementary role to that of the honey bee in pollination. However, their complementary role can only be seen in plantations where their density is sufficiently high, i.e. in orchards located in an environment that allows their populations to survive and develop. It therefore seems essential not to rely on a single pollinator species and to take into account the importance of interactions between natural and cultivated environments for pollinating fauna and for the pollination of both cultivated and wild plants. This is the case of Ophrys pollinated by sexual lures from male Andrena beetles, which feed on the nectar and pollen of other plants, including cultivated species. Farmers need to be made aware of the role played by domestic and wild bees in pollinating crops to improve yields. Many people are aware of the existence of honeybees and the production of honey and wax, but not of the role played by these bees in pollination.Cooperative ventures between growers and beekeepers could be organised, as in California in the western United States (seed producers rent hives from beekeepers). Installing a sufficient number of bee colonies in flowering fields and orchards to ensure adequate pollination of crops not only increases crop yields but also increases honey yields.In order to better understand the complexity of bee-plant relationships, it would be useful to focus future studies on the pollination behaviour of various bee species on other cultivated plants, with a view to comparing the impact of their pollination on yield quality. As part of the improvement of local varieties, it would be desirable to study more closely the bees involved in pollinating these crops to ensure quality yields. Only by working long and hard on several groups of plants and bees will it be possible to propose coherent models. It would also be interesting to supplement this work with a study of the biotic and abiotic factors acting on bee populations. It is also essential to study our apoid fauna. These studies must be extended to cover the whole of the country, including cultivated flora, in order to gain a better understanding of the fauna that pollinate our crops. Observation of the foraging behaviour of these wild species, as well as knowledge of their biology and nesting behaviour, will enable us to select the species best suited to pollinating our crops effectively and to undertake their breeding.

BIBLIOGRAPHICAL REFERENCES

Abdul L., Saeed Ahmad M., Shafqat S., Naeem I., Qamar S. (2019). Diversity of pollinators and their role in the pollination biology of chickpea, Cicer arietinum L. (Fabaceae). Journal of Asia-Pacific Entomology , 22: 597-601.

Abu-Amer J.H., Saoub H.M., Akash M.W., and Al-Abdallat A.M. (2011). Genetic and Phenotypic Variation Among Faba Bean Landraces and Cultivars. International Journal of Vegetable Science, 17: 45-59.

Abrol D.P. (1988). Effect of climatic factors on pollination activity of alfafa pollinating subtropical bees Megachile nana Bingh and Megachile flavipes Spinola (Hymenoptera: Megachilidae). Acta.Oecologica, 9(4): 371-377.

Aguib S. (2006). Etude bioécologique et systématique des Hyménoptères Apoidea dans les milieux naturels et cultivés de la région de Constantine. Magistère thesis in Entomology, Université Mentouri, Constantine, 161p.

Aguib S., Louadi K. & Schwarz M., (2010). The Anthidiini (Megachilidae, Megachilinae) of Algeria with three new species for this country: Anthidium florentinum (Fabricius, 1775), Anthidium amabile (Alfken, 1932) and Pseudoanthidium enslini (Alfken, 1928). Entomofauna, 31: 121-152.

Aguib S., Louadi K., Schwarz M. (2014). The genus Stelis Panzer 1806 (Hymenoptera, Apoidea, Megachilidae) from Algeria with a species new to the fauna of this country. Entomofauna, 35: 553-57.

Alan Walters S. and H.Taylor B. (2006). Effects of honey bee pollination on pumpkin fruit and seed yield. Hort Science , 41(2): 370-373.

Alleaume C. (2012). The honey bee (Apis mellifera), an example for the study of the attractiveness of cultivated plants on pollinating insects. Doctoral thesis in veterinary medicine. École nationale vétérinaire d'Alfort, 102 p.

Aniruddha, M., S.K. Chakrabarty, and J.B. Yadav. (2014). Foraging behaviour of honeybees (Apis spp) (Hymenoptera: Apidae) in hybrid seed production of Indian mustard (Brassica juncea). Indian Journal of Agricultural Science, 84 (11), 1389-94.

Anonymous (2009). Cereal growing. Revue de l'Institut Technique des Grandes Cultures (ITGC), NO. 40.

Anonymous (2010). Technical data sheets on market garden and industrial crops. Lettuce growing. Institut Technique des Cultures Maraichères et Industrielles (ITCMI).

Anonymous (2012). Slight base in citrus production in 2011/2012 (http://www.elmoudjahid.com, Accessed 2019).

Anonymous (2013). Accessed 2019. Apis Bruoc Sella asbl (http://www.apisbruocsella.be/).

Anonymous (2013). Health of bees, Status of situation and role of the Anses http://www.anses.fr/fr/content/ bee-health (ANSES).

Anonymous (2014). Algerian fruit and vegetable production. The essentials of agribusiness and agriculture. Agroligne : n°87.

Anonymous (2017). Vegetable plants (leaves). http://www.dev.scienceenlivre.org/document/plantes-potageres-les-feuilles.pdf.

Aouar-sadli M., Louadi K. and Doumandji S. (2008). Pollination of the broad bean (Vicia faba L.var. major) (Fabaceae) by wild bees and honey bees (Hymenoptera: Apoidea) and its impact on the seed production in the Tizi-Ouzou area (Algeria). African Journal of Agricultural Research, 3 (4) : 266-272.

Aouar-sadli M. (2009). Systematics, eco-ethology of bees (Hymenoptera: Apoidae) and their relationship with field-grown fava beans (Vicia faba L.) in the Tizi-Ouzou region. Doctoral thesis. Universite Mouloud Mammeri de Tizi-Oouzou, Faculté des Sciences biologiques et des Sciences agronomiques, 268p.

Aouar-Sadli M., Louadi K., Doumandji S. (2012). New Records of Wild Bees (Hymenoptera, Apoidea) for Wildlife in Algeria. Journal of the Entomological Research Society (JERS), 14: 19-27.

Arigue S. (2004). L'entomofaune des Hyménoptères Apoidea dans la région saharienne d'El Oued (Djamaâ). Magistère en Entomologie thesis, Univ. Mentouri, Constantine, 102p.

Asif, S., S. Shafqat, and M. Asad. (2008). Pollinator Community of Onion (Allium cepa L.) and its Role in Crop Reproductive Success. Pakitan Journal of Zoology (40): 451- 456.

Asla T., 2002 - Contribution à l'étude de l'efficacité des travaux de D.R.S. en Kabylie (W. de Tizi-Ouzou) - Etat actuel et aspect morphologique des réseaux de banquettes : Use of a survey questionnaire. Magister thesis, Université Mouloud Mammeri de Tizi-Ouzou, 134p.

Avila C.J., Martinho M.R., Campos J.P. De. (1989). Pollination and pollinators in the production of fruits and hybrid seeds of squash (Cucurbita pepo var.melopepo). Annals of the Entomological Society of Brazil,18 (1): 13-19.

Azo'o ela M., Messi J., Tchuenguem Fohouo F., Tamesse J. L., Kekeunou S., Pando J. B. (2010). Foraging behaviour of Apis mellifera adansonii and its impact on pollination, fruit and seed yields of Citrullus lanatus at Nkolbisson (Yaoundé, Cameroon). Cameroon Journal of Experimental Biology, 06 (01): 41-48.

Bakiri A., Louadi K. & Schwarz M. (2016). The genus Nomada Scopoli, 1770 from North - East Algeria (Hymenoptera, Apidae, Nomadini). Entomofauna , 37 : 697-712.

Bakiri E. (2016). Monographie des insectes Hyménoptères Apoidea cleptoparasites en

Algérie. Doctorat 3ème cycle thesis in Animal Biology. Université Frères Mentouri Constantine. 128p

Barbier E. (1986). Crop pollination. Why pollinate? Why? How? Ed Borione, Avignon, 480 p.

Barbier E. (1964). Pollination and fruiting of the Clementine tree. Les Annales de l'Abeille, 7 (1): 63-80.

Batra S.W.T. (1984). Solitary bees. Reproductive behaviour, important role in the pollination of cultivated plants. Pour la Science, 78 : 58 - 68.

Battaglini, M. B. (1969). The importance of honeybees for fertilizing Cucurbita pepo. Apicolt Italia, 35 (1): 9-12.

Belhadi N. (2010). Effet des facteurs d'élevage sur la production et la qualité de lait de vache en région montagneuse. Magister's thesis in agronomy, Université Mouloud Mammeri, Tizi-Ouzou, 83p.

Benachour K., Louadi K. and Terzo M. (2007). Role of wild and domestic bees (Hymenoptera: Apoidea) in the pollination of faba bean (Vicia faba L. var. major) (Fabaceae) in the Constantine region (Algeria). Annales de la société entomologique de France, 43 (2), 213-219.

Benachour K. (2008). Diversity and pollination activity of bees (Hymenoptera: Apoidea) on cultivated plants. Doctoral thesis in applied entomology. Univ. Mentouri, Constantine,151p.

Benachour K. & Louadi K. (2011). Foraging behaviour of bees (Hymenoptera: Apoidea) on male and female flowers of cucumber (Cucumis sativus L.) (Cucurbitaceae) in the Constantine region (Algeria). Annales de la société entomologique de France (n.s.), 47 (1-2): 63-70.

Benarfa N. (2004). Inventaire de la faune apoïdienne dans la région de Tébessa. Thèse de Magistère en Entomologie, Univ. Mentouri, Constantine, 123p.

Bendifallah L. (2002). Biosystématique des Apoidea (abeilles domestiques et sauvages) dans quelques stations de la région orientale de la Mitidja. Magistère en Sciences agronomiques thesis, INA d'Alger, 208 p.

Bendifallah L., Louadi K., Doumandji S. (2010). Apoidea and their diversity in the North from Algeria. Silva Lusitana , 18: 85-102.

Bendifallah L., Louadi K., Doumandji S., Iserbyt S. (2012). Geographical variation in diversity of pollinator bees at natural ecosystem (Algeria). International journal of science and advanced Technology, 2:26 -31

Bendifallah L., Louadi K., Doumandji S. (2013). Bee fauna potential visitors of coriander flowers Coriandrum sativum L. (Apiaceae) in the Mitidja area (Algeria). Verista, Journal of

Apicultural Science, 57:59-70.

Bendifallah L., Koudjil M., Acheuk F., Doumandji S., Louadi K., Boudia I., Achour O. (2015). Spatiotemporal distribution of wild bees across the northwestern regions of Algeria. Revue Nature et Technologie, 12: 86-99.

Benedek, G.E., and E. Gaal (1972). The effect of insect pollination in seed onion with observations on the behaviour of honeybees on the crop. Journal of Apicultural Research, (11), 175-180 B.

Blumler M A., (1991). Modelling the origins of legume domestication and cultivation. Econ. Bot. 45 p.

Borneck, R. and Bricout, J.P. (1984). Evaluation of the economic impact of the pollinating entomofauna in agriculture. Bulletin Technique Apicole, (11): 117-124.

Borneck, R. and Merle, B. (1989). An attempt to evaluate the economic impact of the pollinating bee in European agriculture. Apiacta, (24), 33-38.

Bessaoud O. (2019). Summary report on agriculture in Algeria. Projet d'appui a l'initiative ENPARD Méditerranée, 81p.

Bouri A. and Gaouar S. (2018). Genetic characterization of chickpea accessions (Tlemcen). Contribution à l'analyse génétique et caractérisation de quelques variétés de Pois chiche (Cicer arietinum) au niveau de la wilaya de Tlemcen. Editions universitaires européennes 136p.

Burkle, L.A., Marlin, J.C. and Knight, T.M. (2013). Plant-Pollinator Interactions over 120 Years: Loss of Species, Co-Occurrence, and Function. Science, 339: 1611-1615.

Cane J. H., Sampson Blair J. and. Miller Stephanie A. (2011). Pollination value of male bees: the specialist bee Peponapis pruinosa (Apidae) at summer squash (Cucurbita pepo). Environmental Entomology, 39(5): 000-000.

Chansigaud J. (1975). Study of the behaviour of Andrena carantonica Perez during the flowering of apple trees of the Golden delicious variety. Apidologie, 6 (4) : 341-359.

Chase M. W. and Reveal J. L. (2009). A phylogenetic classification of the land plants to accompany APG III. Botanical Journal of the Linnean Society. vol. 161. p. 122-127

Chedded M.A. (2015). Analyse de l'impact des investissements agricoles réalisés dans le cadre du Plan National de Développement Agricole (PNDA) sur l'évolution des techniques de productions laitières, céréalières et oléicoles en Algérie : Etude de cas dans la wilaya de Tizi-Ouzou. Doctoral thesis in agronomic sciences. Université d'Avignon et des pays de Vaucluse, 153p.

Cherair E.-H., Scheuchl E., Doumandji S., Louadi K. (2013). A new record and a new subspecies of Andrena haemorrhoa (Fabricius, 1781) in Algeria (Hymenoptera: Apoidea

Adrenidae). African Entomology 21: 287- 293.

Cherair El-Hachemi M. (2016). Eco-ethological study of Apoides (Hymenoptera, Aculeata) settlement in steppe environment (Djelfa Region). PhD thesis in agronomic sciences, INA d'Alger: 146 p.

Chichoune H., Benachour K., Louadi K., Javier Ortiz-Sánchez F. (2018). First data on Halictidae (Hymenoptera: Apoidea) from the Batna region (eastern Algeria), Annales de la Société entomologique de France ? 54:1-17.

Collison, C.H. (1973). Nectar secretion and how it affects the activity of honey bees in the pollination of hybrid pickling cucumbers, Cucumis sativus L. M.Sc. Thesis, Michigan State University, East Lansing, USA.

Corbet, S.A., William, I.H. and Osborne, J.L. (1991). Bees and the pollination of crops and wildflowers in the European Community. Bee World, 72: 47-59

Couto R.H.N., Peirera J.M.S., Couto L.A. (1990). Effects of pollination in Cucurbita pepo (summer squash). Cientifica, 18 (1): 21-27.

Crété P. (1965). Précis de botanique tome 2. systématique des angiospermes. Edition Masson et Cie: 238-246

Das Graças Vidal M., Avid De Jong D., Chris Wien H. and A. Morse R. (2010). Pollination and fruit set in pumpkin (Cucurbita pepo) by honey bees. Revista Brasil. Bot. 33(1):107-113.

Delaplane K.S. and Mayer D.F. (2000). Crop pollination by bees. CABI publishing. 344p.

Devika Rani D, Triveni B, Nandini and Narsa Reddy G (2017). Study of Evaluation of Foraging Behaviour of Major Insect Pollinators on Summer Squash (Cucurbita Pepo L.). Journal of Entomology and Zoology Studies, 5(5): 28-30.

Dhaliwal S. M. (2017). Handbook of vegetable crops, Cucurbits (chapter 5). Pp 77-147.

Dibos C. (2011). Plant-pollinator interactions: characterization of the quality of the pollen of two cucurbits during its ontogeny, presentation and transport on the body of the honey bee. Doctoral thesis in agronomic sciences. Université d'Avignon et des pays de Vaucluse, 191p.

Djouama H., Louadi K., Scheuchl E. (2016). Preliminary inventory of the genus Andrena (Hymenoptera: Apoidea, Andrenidae) from some Saharan localities in eastern Algeria. Annales de la Société entomologique de France , 52: 300-310.

Eischen, F.A., (2000). Pollination research and agriculture. Am. Bee J., 140: 118-119.

Esther Julier H. and T'ai H. Roulston (2009). Wild bee abundance and pollination service in cultivated pumpkins: farm management, nesting behavior and landscape effects. Journal of Economic Entomology, 102 (2): 563-573.

Ewies M.A. and EL-Sahhar K. F. (1977). Observations on the behaviour of honeybees on onion and their effects on seed yield. Journal of Apicultural Research, 16 (4), 194-196.

FAO (2019). The importance of bees in biodiversity and their contribution to food security. food and nutritional security.

Feller, C., H. Bleiholder, L. Buhr, H. Hack, M. Hess, R. Klose, U. Meier, R. Stauss, T. van den Boom und E. Weber (1995). Phiinologische Entwicklungsstadien von Gemlisepflanzen. I. Zwiebel-, Wurzel-, Knollenund Blattgemlise. Nachrichtenbl. Deut. Pflanzenschutzd. 47 (8), 193-206

Fluri P., Pickhardt A., Cottier V. and Charrière J. (2001). La pollinisation des plantes à fleurs par les abeilles - Biologie, Écologie, Économie. Agroscope Liebefeld-Posieux, Bee Research Centre, CH-3003 Bern.

Free J.B. (1993). Insect pollination of crops. 2nd ed. Academic Press, London.

Franklin, D.F. (1970). Problems in the production of vegetable seed. In: The indispensable pollinators. Arkansas Agricultural Extension Service Miscellaneous Publications, (127), 112-140.

Garibaldi, L.A., Carvalheiro, L.G., Vaissière, B.E., Gemmill-Herren, B., Hipólito, J., Freitas, B.M. et al. (2016). Mutually beneficial pollinator diversity and crop yield outcomes in small and large farms. Science, 351: 388-391.

Gasim S. M. and Abdelmula A. A. (2018). Impact of Bee Pollination on Yield of Faba Bean (Vicia faba L.) Grown under Semi-Arid Conditions. Agricultural Sciences, 9, 729- 740.

Georges, T., and T.F. Fernand-Nestor (2014). Foraging and pollination activity of Apis mellifera adansonii Latreille (Hymenoptera: Apidae) on flowers of Allium cepa L. (Liliaceae) at Maroua, Cameroon. International Journal of Agronomy and Agricultural Research, (5), 139-153.

Goubara M. and Takasaki T. (2003). Flower visitors of lettuce under field and enclosure conditions. Applied Entomology and Zoology, 38 (4): 571-581.

Goubara M. and Takasaki T. (2003). Pollination effects of the sweat bee Lasioglossum villosulum trichopse (Hymenoptera: Halictidae) on genic male-sterile lettuce. Applied Entomology and Zoology, 39(1): 163-169.

Grajales-Conesa J., Meléndez Ramírez, V., Cruz-López, L. and Sánchez, D. (2013). Native bees in blooming orange (Citrus sinensis) and lemon (C. Limon) orchards in yucatán, mexico. Acta Zoológica Mexicana (n.s.), 29 (2): 437-440.

Greatti M., Zoratti M.I., Trouillier J. (1997). Une autre ressource pour l'apiculteur: l'aide à la pollinisation. De l'abeille de France No. 832.

Hamza N. (2014). Application of arbuscular mycorrhizae in market gardening case of watermelon (Citrullus lanatus). Magister thesis in plant biology and physiology. Université Ferhat Abbas Sétif, 54p.

Hebie I. (2001). Etude experimentale sur la production de semences de nouvelles varietes dans la region de bazega: cas de Lactuca sativa et des hybrides F1 de Hibiscus esculentus et de Solanum melongena. Final dissertation in agronomy, Polytechnic University of Boso-Dioulasso (U.P.B), 47p.

Hodges, L. and Baxendale, F. (1995). Bee pollination of cucurbit crops. Institute of Agriculture and Natural Resources. University of Nebraska-Lincoln.

Hugh c. Cutler and Thomas W. Whitaker (1961). History and distribution of the cultivated cucurbits in the Americas. American Antiquity, 26 (4): 469-485.

Ikhlef H. (2015). Contribution à l'étude systématique et écologique des abeilles sauvages(Hymenoptera: Apoïdea) et l'influence de leur pollinisation sur le rendement du Sulla dans la région de Tizi-Ouzou. Universite Mouloud Mammeri de Tizi-Oouzou, Faculté des Sciences biologiques et des Sciences agronomiques, 108p

INRA. (2005). Measuring the agronomic and economic importance of pollinating insects. INRA Press Office.

Jacob-Remacle, A. (1989 a). Foraging behaviour of the honey bee and of wild bees in apple orchards in Belgium. Apidologie, (20), 271-285.

Jacob-Remacle A., 1989 b - Relations plantes-abeilles solitaires en milieu urbain : l'exemple de la ville de Liège. Comptes rendus du Symposium " Invertébrés " de Belgique : 387-394.

Jacob-Remacle A. (1990). Wild bees and pollination. Unité de Zoologie Générale et Appliquée. Faculté des Sciences Agronomiques de Gembloux, 40 p.

Jaccon G. (2015). FABACEae: Leguminosae, Papilionaceae.(www.gilbertjac.com).

Jean-Marie, P. (1991). Pollination by bees. Placement of colonies in flowering crops with a view to increasing crop yields. EDISUD, Aix-en-Provence.

Jones, H. A. (1927). Pollination and life history studies of lettuce (Lactuca Sativa L.). Hilgardia, 2: 425-448.

Jones, H.A. (1937). Onion improvement, USDA Yearbook 1937, 233-250.

Khelloul L. (2014). Qualitative and quantitative inventory of aphids infesting the fava bean crop. Population dynamics of some characteristic species in two plots of faba bean Vicia faba minor and Vicia faba major in the region of Tizi-Rached (Tizi- Ouzou). Magister's thesis in

Biological Sciences, University of Tizi-Ouzou, 122 p.

Korichi Y. (2015). Contribution to the systematic and eco-ethological study of wild bees (Hymenoptera: Apoïdea) in the Tizi-Ouzou region. Magister thesis. Université Mouloud Mammeri de Tizi-Ouzou. 95 p.

Koltowski Z.(1996). Influence of pollinating insects on yield of several field bean cultivars (Vicia faba L. spp. minor Harz). Pszczelnicze Zeszyty Naukowe 40(1): 77-93.

Krug C, Alves-dos-Santos I. and ames Cane J. (2010). Visiting bees of Cucurbita flowers (Cucurbitaceae) with emphasis on the presence of Peponapis fervens Smith (Eucerini, Apidae) Santa Catarina, Southern Brazil. Oecoologia Australis, 14(1): 128-139.

Kumar Jitender, Mishra R. C. and Gupta J. K. (1985). The effect of mode of pollination on allium species with observation on insects as pollinators. Journal of Apicultural Research, 24(1), 62-66.

Kyllönen T. (2018). Effects of pollination on pod distribution in faba Bean (vicia faba l.). Master's thesis. University of Helsinki, 63 p.

Lachaud A. and Mahé G. (2008). Contribution à la connaissance de la diversité des abeilles sauvages de Loire-Atlantique. Bretagne vivante - SEPNB, Brest, 89p.

Link, W., A.A. Abdelmula, E. Von Kittlitz, S. Bruns, H. Riemer, and D. Stelling. Stelling (1999). Genotypic variation for drought tolerance in Vicia faba L. Plant Breeding, 118: 477-483.

Louadi K., Doumandji S. (1998a). Diversity and foraging activity of bees (Hymenoptera, Apoidea) in a therophyte lawn in Constantine (Algeria). The Canadian Entomologist, 103(5): 691-702.

Louadi K., Doumandji S. (1998b). Note d'information sur l'activité des abeilles (domestiques et sauvages) et l'influence des facteurs climatiques sur les populations. Sciences &Technologies, (9): 83-87.

Louadi K. (1999 a). Systématique, éco-éthologie des abeilles (Hymenoptera, Apoidea) et leurs relations avec l'agrocenose dans la région de Constantine. PhD thesis, Natural Sciences, Université Mentouri, Constantine, 202p.

Louadi K. (1999 b). Contribution à la connaissance des genres Halictus et Lasioglossum de la région de Constantine (Algérie) (Hymenoptera, Apoidea, Halictidae). Bulletin de la Société entomologique de France, 104 (2): 141-144.

Louadi K., Benachour K., Berchi S. (2007a). Floral visitation patterns of bees during spring in Constantine, Algeria. African Entomology, 15(1): 209-213.

Louveaux J., 1980 - Les abeilles et leur élevage. Ed. Hachette, Paris, 230 p.

Maatallah R. (2003). Inventory of the apoidian fauna in the Skikda region. Magistère thesis in Entomology, Univ. Mentouri, Constantine, 172p.

Maghni N. (2006). Contribution à la connaissance des abeilles sauvages (Hymenoptera: Apoidea) dans les mileux naturels et cultivés de la région de khenchela. Magisterial thesis in Entomology, Univ. Mentouri, Constantine, 127 p.

Maghni N, Louadi K, Ortiz-Sanchez FJ, Rasmont P. (2017). The Anthophores of the Aurès region, northeastern Algeria (Hymenoptera: Apidae: Anthophorini). Annales de la Société entomologique de France, 53: 55-73.

Malerbo-Souza D. T., Nogueira-Couto R. H., Couto L. A. (2004). honey bee attractants and pollination in sweet orange, Citrus sinensis L. OSBECK, VAR. PERA-RIO.

Marzinzig Birgit, Brünjes Lisa, Biagioni Siria, Behling Hermann, Link Wolfgang and Westphal Catrin (2018). Bee pollinators of faba bean (Vicia faba L.) differ in their foraging behavior and pollination efficiency. Agriculture, Ecosystems and Environment, 264: 24-33.

Mc Gregor S.E. (1976). Insect pollination of cultivated crops plants. US Department of Agriculture, Agriculture Handbook No 496, Washington, 411 p.

Mesquida J., Le Guen J., Tasei Jn., Carre S. and Morin G. (1990). Modalities of pollination in two lines of spring faba bean (Vicia faba L. var equina Steudel). Effects on castings, productivity and crossing rates. Apidologie, 21, 511-525.

Michelbacher, A. E., Ray F. Smith, and P. D. Hurd, Jr (1964). Bees are essential pollination of squash, gourds, and pumpkins. California Agriculture, 18: 2-4.

Michelbacher A.E., Smith Ray F., Hurd P.D. Jr. 1964. Bees are essential... Pollination of Squash, Gourds, and pumpkins. Calif ornia Agriculture, 18: 2-4.

Michener C.D. (2000). The Bees of the World. The Johns Hopkins University Press, 807p.

Mortaza, R.A., H. Sadeghi-Namaghi, and H. Mojtaba (2013). Effects of insect pollinators on onion seed production quality and quantity. Journal of Crop Protection. (4): 395-402.

Mudssar A., Shafqat S., Asif S and Muhammad Amjad B.(2014). Exploring the best native pollinators for pumpkin (Cucurbita pepo) production in Punjab, Pakistan. Pakistan Journal of Zoology, 46(2): 531-539.

Musallam Iyad W., Haddad Nizar J., Tawaha Abdel-rahman M. and Migdadi Osama S. (2004). The Importance of Bee-Pollination in Four Genotypes of Faba Bean (Vicia faba L.). International journal of agriculture & biology, 6(1): 9-12.

Nepi M., Pacini E. (1993). Pollination, Pollen Viability and Pistil Receptivity in Cucurbita pepo. Annals of Botany, 72: 527-536.

Nevkryta, A. N. (1953). Insects pollinating cucurbit crops. 92 pp. [Kiev] Akad. Nauk Ukrain. SSR. AA153/61.

Nieto, A., Roberts, S.P.M., Kemp, J., Rasmont, P., Kuhlmann, M., García Criado, M., et al. (2014). European Red List of Bees. Luxembourg: Publication Office of the European Union.

Njoroge G.N.,G emmill B., Bussmann R., Newton L.E. and Ngumi V.W. (2004). Pollination ecology of Citrullus lanatus at Yatta, Kenya. International Journal of Tropical Insect Science, 24 (1): 73-77.

Njoroge G.N., Gemmill B., Bussmann R., Newton L.E. and Ngumi V.M. (2010). Diversity and efficiency of wild pollinators of watermelon (Citrullus lanatus (Thunb.) Mansf.) at Yatta (Kenya). Journal of Applied Horticulture, 12(1): 35-41.

Nye, W. P., N.S. Shasha'a, W.F. Campbell, and A.R. Hamson (1973). Insect Pollination and Seed Set of Onions (Allium cepa L.). Pollinating Insects Research Unit.

Praloran C., 1971 - Les agrumes. Ed. editeur 8348, Paris, n° 5, p. 25.

Patiny S. and Terzo M., (2010). Catalogue and key to the subgenera and species of the genus Andrena from Belgium and northern France (Hymenoptera, Apoidea).Université de Mons Laboratoire de Zoologie B-7000 Mons, 39 p.

Pauly A., (2014). Key to the Halictidae of Belgium. Atlas Hymenoptera working paper.

Payette A. (1996). The Apoids of Quebec: Bees and agriculture. L'Abeille, 16 (4): 14-15.

Pesson P, Louveaux J (1984). Pollination and plant production. INRA, Paris, pp. 663.

Pfiffer L. and Mûller A. (2014). Wild bees and pollination. Institute of FiBL (Research Foundation for Organic Agriculture).

Pierre J., Le Guen J. Pham Delègue MH., Mesquida J., Marilleau R., Morin J. (1996). Comparative study of nectar secretion and attractivity to bees of two lines of spring- type faba bean (Vicia faba L. var. equine Steudel). Apidologie, (27) : 65-75.

Pierre J, Le Guen J, Esnault R, Debbagh S, Sadiki M (1997). Méthode d'étude de la fréquentation de diverses féveroles par les insectes pollinisateurs, p. 199-206 in: INRA (ed.) Les légumineuses alimentaires méditerranéennes. Rennes (France), 20-22 February, Les Colloques, 88, INRA, Paris.

Pierre J, Suzo MJ, Moreno MT, Esnault R, Le Guen J (1999). Diversity and efficiency of the pollinating entomofauna (Hymenoptera: Apidae) of faba bean (Vicia faba L.) at two sites in France and Spain. Annales de la société entomologique de France (n.s.), 35 : 312-318.

Poole, C. F. and Porter D.R. (1933). Pollen germination and development in watermelon. American Society for Horticultural Science,30: 526-530.

Pouvreau A. (2004). Pollinating insects. Delachaux & Niestlé, 157 p.

Rao, G.M. and M.C. Suryanarayana, (1988). Studies on pollination of watermelon (Citrullus lanatus (Thunb.) Mansf.). Indian Bee Journal, 50: 5-8.

Rasmont, P., J.Leclercq, A.Jacob-Remacle, A.Pauly & C.Gaspar, (1993) The faunistic drift of Apoidea in Belgium. pp. 65-87 in E. Bruneau, Bees for pollination, Commission of the European Communities, Brussels, 237 p.

Rasmont P., Y. Barbier, S. Iserbyt, R. Wahis & M. Terzo (2003). Could Jean-Henry Fabre observe all these insects today? pp. 209-220 in Jean-Henri Fabre, un autre regard sur l'insecte. Proceedings of the International Conference on Entomology, 18-19 October 2002. Saint-Léons en Lévézou (France, Aveyron). Conseil général de l'Aveyron, Rodez, 275 p.

Rasmont P., and Terzo M., (2010). Catalogue and key to the subgenera and species of the genus Bombus from Belgium and northern France (Hymenoptera, Apoidea). Université de Mons Laboratoire de Zoologie. 28p.Ribeiro Sousa G., Alves E., Lopes De Carvalho C. A. (2017). Biology of pollination of Citrus sinensis variety 'PERA RIO'. Revista Brasileira de Fruticultura, 39 (2), 1-8.

Rasmont P. & Iserbyt S. (2012). The Bumblebees Scarcity Syndrome: Are heat waves leading to local extinctions of bumblebees (Hymenoptera: Apidae: Bombus)?. Annales de la Société entomologique de France, (N.S.) 48(3-4: 275-280.

Reta Sanchez D.G., Santos Serrato Corona J., Viramontes R.F., Cueto Wong J.A., Padilla S.B., César J.S. (2008). Cultivos alternativos con potencial de uso forrajero en la comarca lagunera, Primera, Mexico, pp 41

Richard K.W. and Edwards P.D., (1988). Density, Diversity and efficiency pollinators of sainfoin, Onobrychis viciaefolia Scop. The Canadian Entomologist, 120 (12): 1085-1100.

Robinson, W.S., Nowogrodzki, R. and Morse, R.A. (1989) The value of honey bees as pollinators of U.S. crops. American Bee Journal, 129: 411-423, 477-487.

Sanduleac, e. (1959). Data on the entomophilous pollination and the selection Of Cucurbitaceae. Lucr. Stiint. Stat. Cent. Seri. Apic. 1: 129-132. [Aa-431/61.

Sanford, M.T. (1992). Bee keeping: Watermelon pollination. Co-operative extension service RF-AA091, Florida.

Schweisguth, B. (1976). Onion physiology, cultivar adaptation and storability. INVUFLEC. France, pp. 27-32.

Sedgley M. and Buttrose M.S. (1978). Some effects of light intensity, daylength and temperature on flowers and pollen tube growth in the watermelon (Citrullus lanatus). Annals of Botany, 42: 609-616.

Shebl Mohamed A. & Farag Mahmoud (2015). Bee diversity (Hymenoptera: Apoidea) visiting Broad Bean (Vicia faba L.) flowers in Egypt. Zoology in the Middle East, 61(3): 256-263.

Selka O. (2007). Study of infestations o f Citrus sinensis var. San guinelli by Ceratitis capitata (diptera: tephritidae) and by the pathogenic microflora in the plain of hennaya (Tlemcen). Magister thesis in animal ecology. University of Abou bekr Belkaid-Tlemcen.

Saxena M.C. (1991). Status and scope for production of faba bean in the Mediterranean countries. Mediterranean Options. Seminar Series 10: 15-20.

Schweisguth B. (1976). Physiology of the onion, adaptation of cultivars and aptitude for conservation. Institut National de Vulgarisation pour les Fruits, Légumes et Champignons (INVUFLEC). France, pp 27-32.

Singh A. K. (2016). Insect pollinators of sweet orange and their attributes. Indian Journal of Entomology, 78(1): 46-50.

Somerville Doug (2002). Honeybees in faba bean pollination.NSW Agriculture Agnote DAI-128.

Sonnet M., Jacob-Remacle A. (1987). Pollination of the forage legume Hedysarum coronarium L. in Tunisia. Bulletin des Recherches Agronomiques de Gembloux, Belgium, 22(1): 19-32.

Stanghellini, M.S., J.T. Ambrose and Schultheis J.R., (2002). Diurnal activity, floral visitation and pollen deposition by honey bees and bumblebees on fi eld-grown cucumber and watermelon. Journal of Apicultura. Research, 41: 27-34

Stoddard FL, Bond DA (1987). The pollination requirements of the faba bean. Bee World, 68(3): 144-152.

Stone, G.N., P. Willmer, and J.A. Rowe (1998). Partitioning of pollinators during flowering in an African acacia community. Ecology, 79 (8): 2808-2827.

Suzo M.J., Pierre J., Moreno M.T., Esnault R., Le Guen J. (2001). Variation in outcrossing levels in faba bean cultivars: role of ecological factors. Journal of Agricultural Science, 136: 399-405.

Taha, El-Kazafy A, and Yousry A B. n.d. (2009). The Value of Honey Bees (Apis Mellifera, L.) as Pollinators of Summer Seed Watermelon (Citrullus Lanatus Colothynthoides L.) in Egypt. Acta Biologica Szegediensis, 53(1) : 33-37.

Tasei J.N. (1976). Insect pollinators of winter faba bean (Vicia faba equina l.) and pollination of male-sterile plants in hybrid seed production. Apidologie 7(1): 1-38.

Tepedino V.J. (1981). The pollination efficiency of the squash bee (Peponapis pruinosa) and the honey bee (Apis mellifera) on summer squash (cucurnita pepo). Journal of the Kansas Entomological Society, 54(2): 359-377.
Terzo M. and Rasmont M. (2010). Key to the genera of Western European apoids. Laboratoire de Zoologie, Université de Mons-Hainaut, pp. 23- 59.

Thompson, R. C. (1933). Natural cross-pollination in lettuce. American Society for Horticultural Science Proc. 30: 545 - 547.

Vaissière B. E. (2002). Bees and pollination. Le courrier de la Nature **196**, Spécial Abeilles: 24-27.
Vaissière B. E. (2005). Bees, pollination and pesticides. Copyright Académie d'Agriculture de France.
Vansell H. and Watkins,G.(1942). Orange nectar and pollen in relation to bee Activity. Journal of economic entomology, 35(3): 321-323.
Varis Anna-Liisa and Brax Raija (1990). Effect of bee pollination on yield and yield components of field bean (Vicia faba L.). Journal of agricultural science in finland, Maataloustieteellinen Aikakauskirja, 62: 45-49.

Vidal M., De Jong D, WIEN H. C. and Morse R. A. (2006). Nectar and pollen production in pumpkin (Cucurbita pepo L.). Revista Brasileira de Botânia, 29(2) : 267-273.

Vidal M., De Jong D, Wien H. C. and Morse R. A. (2010). Pollination and fruit set in pumpkin (Cucurbita pepo) by honey bees. Revista Brasileira de Botânica, 33 (1): 107-113.
Vilain J. (1997). Pollination of courgette. Fiche technique. Groupement Regional des Apiculteurs Pollinisateurs Professionnels "Mediterranéen".
Wadlow R. V. (1970). Pollination of crops in Florida, pp. 61-63, in: The Indispensable Pollinators, Publications of the Michigan Agricultural Extension service 127.
Wehner T(2008). Watermelon in Vegetables I Asteraceae, Brassicaceae, Chenopodicaceae and Cucurbitaceae. Handbook of plant breeding volume I.

Wills D.A., Wearing A.H. (1993). The effect of honeybee pollination on yield and fruit quality in zucchini (Cucurbita pepo L). Australasian Postharvest Conference: 373.
Wolfenbarger D.D. (1962). Honey bees increase squash yields, Florida Agric. Experiment station Research Report, 7(1): 15-19.
Zdzislaw, W., J. Giejdasz, and G. Proszynski (2004). Effect of pollination on onion seeds under isolation by the mason bee (Osmia rufa L.) (Apoidea, Megachilidae) on the setting and quality of obtained seeds. Journal of Apicultural Science, (4), 35-41.
Zorrig W. (2011). Research and characterization of determinants controlling cadmium accumulation in lettuce "Lactuca sativa". PhD thesis in biological sciences, Université Tunis El Manar, 246p.

APPENDICES

Appendix 1. Monthly climate data for the wilaya of Tizi-Ouzou (period 2007- 2017) (data from the Boukhalfa meteorological office).

Average monthly temperatures

	I	II	III	IV	V	VI	VII	VIII	IX	X	XI	XII
2007	11.4	13.0	12.3	15.7	19.5	23.3	27.5	28.1	23.7	19.4	13.6	10.6
2008	11.0	12.6	13.0	16.3	18.8	23.5	27.9	27.8	24.4	19.8	13.5	13.1
2009	10.8	10.4	12.7	14.0	20.8	25.2	29.6	28.0	22.8	19.3	15.3	15.8
2010	11.4	12.8	13.7	16.4	17.6	22.6	27.7	27.2	23.6	19.2	14.3	11.2
2011	10.6	10.3	13.4	17.6	19.7	23.2	27.5	28.5	24.8	19.6	15.7	11.7
2012	9.9	6.5	13.3	15.3	19.9	26.4	27.8	30.3	24.4	20.8	16.3	11.9
2013	10.8	9.4	14.3	15.8	17.5	22.0	26.7	26.9	24.1	23.1	13.6	11.2
2014	12.3	12.8	12.3	17.5	19.2	23.7	27.0	28.0	26.2	21.6	17.3	10.9
2015	9.9	9.8	13.1	17.3	21.4	24.7	29.7	28.8	24.4	20.2	14.9	12.2
2016	12.8	12.6	12.5	16.1	19.0	24.3	27.9	27.1	24.3	22.5	15.7	12.6
2017	8.9	12.7	14.5	16.4	21.8	26.4	29.4	29.7	23.9	19.5	13.7	10.4
2018	11.2	9.9	13.6	16.5	17.8	22.8	27.8	27.4	25.3	19.1	14.9	11.9

Rainfall (mm)

	I	II	III	IV	V	VI	VII	VIII	IX	X	XI	XII
2007	8.4	63.4	284.6	201.5	23.8	14.9	11.3	1.8	56.9	210.4	257.8	89.2
2008	8.7	11.8	88.1	32.3	96.8	14.0	7.2	0	34.6	66.1	190.3	115.4
2009	221.0	42.4	92.0	132.6	69.1	0.0	2.6	7.3	169.8	40.4	116.5	140.3
2010	82.4	61.1	97.4	93.6	59.3	27.6	1.8	25.9	25.4	113.4	143.5	76.1
2011	90.9	146.9	89.9	106.5	153.1	41.1	1.7	0.2	7.5	34.1	156.6	120.6
2012	69.5	269.5	97.8	146.8	40.2	1.1	0.0	6.4	10.9	96.3	68.7	39.8
2013	211.8	185.7	93.5	64.5	151.6	0.0	0.3	11.3	37.6	39.0	164.9	103.7
2014	110.1	110.2	172.4	5.3	10.0	48.4	0.2	3.6	11.8	26.5	61.6	272.4
2015	200.9	181.5	70.2	0.0	10.4	15.3	4.1	2.6	41.1	81.7	102.7	0.0
2016	59.9	96.2	185.3	61.8	68.4	5.9	0.0	0.0	4.2	16.6	68.4	150.1
2017	250	36	29	37.0	2	8.8	0.4	0	37.8	35.3	126	138.0
2018	31	108	187	102.0	33	34	0.5	0	42	182	175	44.0

Average humidity (%)

	I	II	III	IV	V	VI	VII	VIII	IX	X	XI	XII
2007	80	79	98	84	72	69	56	59	67	77	81	86
2008	83	79	78	70	78	68	60	59	65	77	78	83
2009	60	47	48	54	42	29	31	35	49	49	51	58
2010	80	77	78	80	73	68	59	62	70	74	79	76
2011	84	83	78	76	77	75	68	68	69	75	81	85
2012	84	88	81	76	71	61	60	49	65	69	79	83
2013	80	80	73	77	78	65	64	60	73	69	83	83
2014	78	76	81	67	68	64	57	58	62	68	69	84
2015	82	80	75	71	62	60	51	59	65	72	80	75
2016	76	73	76	75	71	59	55	59	62	65	72	85
2017	82	75	72	68	62	58	49	56	61	66	76	83
2018	79	80	75	75	80	68	59	60	67	76	77	83

Average wind speed (m/s)

	I	II	III	IV	V	VI	VII	VIII	IX	X	XI	XII
2007	1.0	2.3	2.2	1.6	1.6	2.3	2.2	2.0	1.4	1.2	1.0	0.3
2008	0.6	0.6	2.3	1.5	1.4	1.1	2.3	2.3	1.5	1.0	1.2	0.8
2009	1.5	1.3	1.4	1.1	1.2	1.8	1.8	1.7	1.1	0.7	0.5	1.2
2010	2.1	1.5	1.1	0.8	1.0	1.3	1.8	1.3	1.4	1.0	0.9	0.9
2011	0.2	1.2	0.7	1.1	xx	xx	xx	xx	1.1	0.5	0.6	0.3
2012	0.3	xx	xx	xx	xx	xx	xx	xx	0.7	0.6	1.1	0.9
2013	1.5	1.6	2.6	1.9	1.9	1.7	1.8	1.3	1.3	xx	xx	xx
2014	xx	xx	xx	xx	xx	xx	xx	xx	xx	xx	xx	xx
2015	1.0	2.0	1.3	1.7	xx	xx	xx	xx	2.8	2.3	1.7	1.4
2016	2.2	2.5	2.5	2.1	2.7	2.5	2.8	2.5	1.8	1.7	1.5	0.9
2017	1.5	2.2	2.1	2.2	2.2	2.7	2.4	2.9	2	1.2	1.0	1.0
2018	1.5	0.8	1.9	1.2	0.7	0.3	1.2	1.9	1.6	1.3	1.3	0.8

Sunshine duration (hours)

	I	II	III	IV	V	VI	VII	VIII	IX	X	XI	XII
2007	195	168	197	154	261	296	353	312	233	182	191	141
2008	182	187	221	262	193	294	300	315	172	101	182	139
2009	108	170	239	197	104	98	111	101	72	241	183	106
2010	140	109	180	200	240	311	368	317	267	209	114	142
2011	142	156	211	243	244	291	356	341	282	233	134	145
2012	201	155	207	210	309	298	344	333	237	232	151	172
2013	140	126	161	231	241	352	335	335	237	210	120	163
2014	161	164	181	278	320	313	360	333	244	256	150	122
2015	143	114	221	297	307	354	395	321	249	211	191	215
2016	168	159	182	217	276	347	377	366	260	234	171	140
2017	133	154	273	270	312	341	349	307	295	271	170	127
2018	203	140	155	199	212	340	325	336	225	196	171	187

SUMMARY

Investigations carried out in 2016, 2017 and 2018 in the Tizi-ouzou region on the Apoids that pollinate six market garden plants and one fruit tree, the orange tree, revealed that the honey bee is the main pollinator of courgettes, watermelons and orange trees. This bee is also the main visitor to watermelon, alongside four other solitary bee species (Nomada sp. Ceratina saundersi, Ceratina cucurbitina, and Lasioglossum pauxillum). The two Halictidae L. pauxillum and L. malachurum appear to be the main pollinators of lettuce and onions. On the other hand, the two legumes, fava beans and chickpeas, are mainly pollinated by other species of the Apidae family: Eucera numida, Eucera decolorata and Eucera eucnemidea. The Megachilidae Lithurgus chrysurus is also the most frequent visitor of chickpeas. Our results also show that the bees' foraging activity follows the anthesis of the flowers studied. It begins as soon as the first flowers open and intensifies as the number of open flowers increases. The study of the foraging behaviour and efficiency of the most abundant bees on the flowers revealed that all the bees' visits can be fertilising except in the case of the faba bean, where 72% of the honey bees' visits to collect nectar are not fertilising (negative foraging). In addition, a study of the effect of entomophilic pollination on bean and onion production revealed that the presence of pollinating insects significantly improves yield. In the case of broad beans, the average weight of seeds harvested from free quadrats was three times higher than that from caged quadrats. As for onions, the presence of foraging bees considerably improved yield, with an economic advantage of around 98% in terms of average seed weight per quadrat and per plant.

Key words : Wild bees, honey bees, foraging, pollination, market garden plants, orange trees, performance, Tizi-Ouzou

MIX
Papier aus verantwortungsvollen Quellen
Paper from responsible sources
FSC® C105338

Printed by Books on Demand GmbH, Norderstedt / Germany